Student Study Guide

for use with

Foundations in Microbiology

Fifth Edition

Kathleen Park Talaro

Williams College

Prepared by

Nancy Boury
Iowa State University

 **Higher Education**

Boston   Burr Ridge, IL   Dubuque, IA   Madison, WI   New York   San Francisco   St. Louis
Bangkok   Bogotá   Caracas   Kuala Lumpur   Lisbon   London   Madrid   Mexico City
Milan   Montreal   New Delhi   Santiago   Seoul   Singapore   Sydney   Taipei   Toronto

The McGraw·Hill Companies

Student Study Guide for use with
FOUNDATIONS IN MICROBIOLOGY, FIFTH EDITION
KATHLEEN PARK TALARO

Published by McGraw-Hill Higher Education, an imprint of The McGraw-Hill Companies, Inc.,
1221 Avenue of the Americas, New York, NY 10020. Copyright © 2005, 2002, 1999 by
The McGraw-Hill Companies, Inc. All rights reserved.

2 3 4 5 6 7 8 9 0 QPD QPD 0 9 8 7 6 5

ISBN 0-07-255303-0

www.mhhe.com

Introduction

Welcome to the wonderful world of microbiology! You are well on your way to learning more about the millions of microscopic inhabitants that we share this planet with. The Foundations of Microbiology text is a great place to start gathering more information. This study guide was written with students in mind, as a tool to help you focus your efforts and make the most of your study time.

Microbiology, like any discipline, has a vocabulary all its own. It has been said that there are more terms in the average beginning biology class than in a first-semester foreign language class. In order to communicate microbiological information effectively with instructors, classmates and future colleagues, you must first learn the language. One of the best ways to learn this new language is to increase your exposure to it. Attend class, read your text and form study groups. With this study guide, answer all the concept questions in your own terms first, then overlay the specialized terminology of the microbiologist.

Knowledge of terminology is very important, but not adequate as an end onto itself. College level courses will require you to possess higher-order thinking skills. Comprehension and application skills will enable you to move beyond the *memorize-and-regurgitate* methods of studying that you may be using. If you can apply the material you are studying you will be able to compare and contrast related and unrelated concepts. You will also be able to generalize trends based on specific examples, or make educated guesses about specific examples based on known trends. These are the skills that allow you to answer "what if" questions and to troubleshoot problems. These are also the skills that will help you succeed both in class and out.

I have written this study guide with both your, the students', and my fellow instructors expectations in mind. I have taught both entry-level and senior-level science courses at the college and university settings for several years. Each year I prepare study guides based on the text I'm using to help my students focus their efforts and make more efficient use of their study time. I also try to help students organize the information presented in the text so they can remember it and apply it to more complex problems. With this study guide I have separated each chapter into three sections:

I. Building Your Knowledge
- *Highlighting key concepts*
- *Review material from the text*
- *Provides an opportunity to use the terminology discussed in the text*

II. Organizing Your Knowledge
- *Provides structure and a framework for the material*
- *Reviews the basics of each chapter*
- *Additional practice with using the vocabulary of microbiology*
- *Compares and contrasts several related topics*

III. Practicing Your Knowledge
- *Gives sample exam questions and tests your preparedness*
- *Most helpful if completed without notes or text to reference*

I would like to thank my husband and the rest of my family for their patience and help as I've worked my way through this project. I would also like to thank the students in my first-year and senior microbiology courses for their suggestions and even their constructive criticism. I will leave you with the advice I give to my students each semester, "Your study guide should be your best friend, not a passing acquaintance." Enjoy!

Warm Regards,
Dr. Nancy Maroushek Boury

Table of Contents

Chapter 1 The Main Themes of Microbiology

Building Your Knowledge

1) Describe the major branches of microbiology. Can a microbiologist study more than one branch of microbiology at the same time? Explain your answer.

2) Where have bacteria been cultured from?

   Were eucaryotes or procaryotes the first living things on the planet?

   Are procaryotes or eucaryotes the most numerous organisms on the planet?

3) How are photosynthesis and decomposition related to one another?

4) What is genetic engineering?

   Name several compounds that microbes have been used to make?

5) Bioremediation utilizes microbes to do what task?

6) What are pathogens?

7) How are procaryotes and eucaryotes similar to one another, but different from viruses?

8) How are procaryotes and eucaryotes different from one another?

9) How many variables can be tested at the same time in an experiment? Why?

10) Joseph Lister made a huge contribution to medicine – what was it?

11) Describe the germ theory of disease. Which two scientists are considered the "founders of microbiology"?

12) What are the 3 domains of life? Which are procaryotic? How does this differ from the 5 Kingdom system?

Organizing Your Knowledge

Please fill in the blanks and boxes, using your text as a reference.

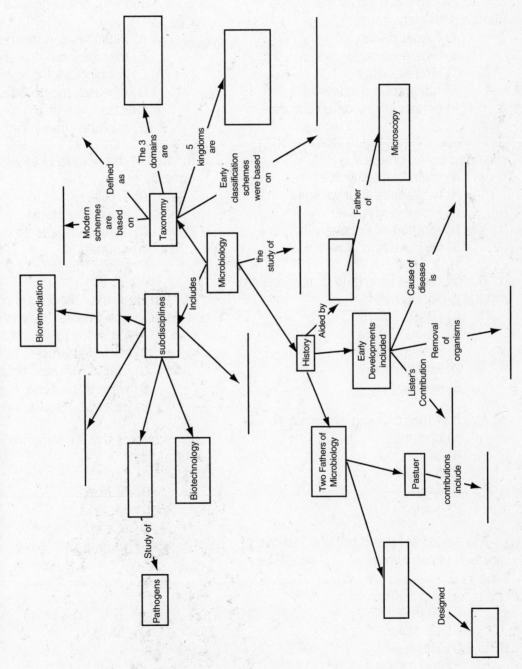

Practicing Your Knowledge

1. _____ monitor and try to control the spread of diseases in communities.
   A) Genetic Engineers
   B) Industrial microbiologists
   C) Virologists
   D) Immunologists
   E) Epidemiologists

2. Microbes that are newly discovered human pathogens cause _____.
   A) fungal diseases
   B) emerging diseases
   C) viral diseases
   D) ubiquitous diseases
   E) genetically modified diseases

3. In modern times organisms are classified based on their _____.
   A) structural similarities
   B) similarities in physiology
   C) morphology
   D) location of discovery
   E) similarities in genetics

4. The process of arranging organisms into orderly groups is called ___.
   A) classification
   B) nomenclature
   C) bioremediation
   D) identification
   E) serology

5. Which of the following are procaryotic?
   A) bacteria
   B) viruses
   C) fungus
   D) algae
   E) humans

6. The worldwide death toll of infectious disease is approximately ____ people per year.
   A) 1 million
   B) 80 million
   C) 13 million
   D) 1 billion
   E) 200 million

7. Ancient procaryotes added _____ to an atmosphere that had very little ____.
   A) water : nitrogen
   B) oxygen : oxygen
   C) carbon: water
   D) carbon: oxygen
   E) hydrogen: hydrogen

8. The germ theory of disease states that ____.
   A) microbes can form spores
   B) microbes are procaryotic
   C) antibiotics kill bacteria
   D) diseases may be caused by infection
   E) microbes may infect abiotically

9. Most of the world's photosynthesis is done by _____.
   A) Trees
   B) Flowering plants
   C) Agricultural plants
   D) Animals
   E) Microorganisms

10. Which of the following is correctly ordered, from smallest to largest?
   A) viruses, fungus, bacteria
   B) bacteria, viruses, fungus
   C) fungus, viruses, bacteria
   D) viruses, bacteria, fungus
   E) bacteria, fungus, viruses

11. Which of the following may be pathogens?
   A) Viruses
   B) Yeasts
   C) Bacteria
   D) All of the above
   E) None of the above

12. Antonie van Leeuwenhoek was a pioneer in the field of ____.
   A) aseptic technique
   B) sterilization
   C) microscopy
   D) spontaneous generation

13. Which of the following are procaryotic domains?
   A) Archaea and Monera
   B) Bacteria and Archaea
   C) Monera and Eukarya
   D) Eukarya and Archaea
   E) Monera and Bacteria

14. A good hypothesis must be supported or discredited by _____
   A) careful thought
   B) repeated inferences
   C) observations or experiments
   D) popular opinion
   E) ancient theories

15. When designing experiments, scientists use control groups to____.
   A) compare against the test groups
   B) test multiple variables at the same time
   C) test five or more variables
   D) support spontaneous generation
   E) all of the above are correct

Chapter 2 From Atoms to Cells

Building Your Knowledge

1) What is the smallest piece of an element that still maintains the properties of that element?

2) Draw a helium (He) atom, labeling protons, neutrons and electrons.

3) What is the atomic number of an element?

How is it different from the atomic mass?

If we add 2 electrons to an element, do we change its atomic mass?

Does this change its atomic number? Explain

If an uncharged atom has 12 protons, how many electrons does it have?

4) What are elements that have the same atomic number, but different atomic masses called?

5) Where are electrons located in an atom? How are they arranged?

What are valence electrons?

6) Which part(s) of an atom are lost, gained or shared during chemical bonding?

7) If an atom has 6 valence electrons and the valence shell can hold 8 electrons, how many electrons can it accept?

How many bonds can it participate in?

8) Compare and contrast ionic and covalent bonds.

Where are electrons given and taken?

Where are the electrons shared?

9) Name one polar compound. Most lipids are non-polar, what does that mean?

10) How do cations differ from anions? How are cations and anions similar?

11) How does hydrogen bonding differ from ionic bonding?

12) Differentiate between molecular and structural chemical formulas.

Do fructose and glucose have the same molecular formula?

Do they have the same structural formulas?

13) Write a chemical reaction combining molecule A with B to make compound C. Label the products and the reactions. Is this a synthesis, decomposition, or exchange reaction?

14) You make a glass of lemonade from powder. What is your solvent?

What is your solute and what do you call your solution?

15) What are amphipathic molecules?

Are hydrophobic molecules polar or non-polar?

Which molecules (polar or non polar) dissolve easily in water.

16) If you dissolve 5 grams of salt in 100 mL of water, what is the concentration of the solution (in %)?

If you dissolve 5 grams of the same salt in 50 mL of water, is the concentration higher or lower than the first solution?

17) An acid releases _____ when dissolved in water. A base releases _____.

If a solution has a pH of 3, does it have more or less $H+$ ions than a solution with a pH of 5?

Is the pH of 3 solution more or less acidic than the pH of 5 solution?

18) Why is sugar ($C_6H_{12}O_6$) considered an organic compound, but carbon dioxide (CO_2) considered an inorganic compound?

19) Why is carbon considered a fundamental element of life?

How many bonds can a single carbon atom participate in?

What types of bonds can a carbon atom participate in?

20) What are functional groups? What is the difference between a hydroxyl and a carboxyl? What molecules have phosphates in them?

21) What are the four macromolecules commonly found in living systems?

     a. _____

     b. _____

     c. _____

     d. _____

22) If you discover a new compound and call it newbose, you are telling the world this is a -

23) What is dehydration synthesis and how is it related to the building of carbohydrates? How are carbohydrates broken down?

24) Why don't oil and water mix?

What are lipids used for in cells and why is it important that they don't dissolve in water?

25) What subunits combine to form lipids and how are the lipids held together?

What do lipases do to lipids?

26) Draw a cell membrane, labeling the phospholipids (head and tail), proteins, hydrophobic regions and hydrophilic regions.

27) Why is it important for membranes to be selectively permeable? What would happen if they were not permeable or were permeable to everything?

28) Draw an amino acid, labeling the amino group, the carboxyl group and the R group. How are amino acids linked together?

29) Are peptide bonds ionic, covalent or hydrogen bonds?

30) Describe four separate functions proteins have in cells.

31) What are DNA and RNA made of? List 2 ways DNA is different from RNA.

32) Draw a nucleotide. Labeling the nitrogen base, pentose sugar and the phosphate group.

33) What is the difference between a purine and a pyrimidine? Which bases are purines? Which are pyrimidines?

34) The sugar-phosphate backbone of a DNA molecule is held together by _____ bonds. The nitrogen base "rungs" are of the DNA ladder are held together by _____ bonds.

35) What are the 3 major types of RNA? What is the function of each?

36) What is ATP? What is it used for in cells? How is it similar to RNA and DNA?

37) Is binary fission a form of sexual or asexual reproduction?

38) List the six properties that define life. Do viruses fulfill each of the six properties? Are viruses alive?

Organizing Your Knowledge

Macromolecule	Subunits	Use	Examples
Carbohydrates		Storage	
			Cellulose
	Fatty Acids & Glycerol		
Proteins			Enzymes
	Nucleotides		

Carbohydrate	Source	Use	Molecular structure
Agar			Sulfur-conjugated carbohydrates
	Fungus & Arthropods	Exoskeletons	
Peptidoglycan		Cell Walls	
	Gram negative cell walls	Structure (causes fever)	
Cellulose			Fibrous long-chains of carbohydrates
	Bacteria, other cells	Attachment	
Starch or Glycogen			long chain carbohydrates

Level of Protein Structure	Description
Quaternary	
	The order of amino acids in a protein chain
Secondary	
	Bonds between functional groups (e.g. disulfide bonds)

Lipid	Common Function	Structure
Phospholipids		Amphipathic
	reinforces cell membranes	Rings (steroid)
Prostaglandins		

Function	Trait	Procaryotic Cells (YES/NO/SOME)	Eucaryotic Cells (YES/NO/SOME)
Reproduction	Mitosis		
	Sex Cells		
Biosynthesis	Golgi Apparatus		
	Ribosomes		
	Endoplasmic Reticulum		
Shape/Protection	Cell Walls		
	Capsules		
Genetics	Nucleic acids		
	Chromosomes		
	Nucleus		
Motility	Flagella		
	Cilia		
Respiration	Enzymes		
	Mitochondria		
Photosynthesis	Pigments		
	Chloroplasts		

Practicing Your Knowledge

1. Hydrophobic molecules are _____ and _____ dissolve easily in water.
   A) nonpolar:do not
   B) polar: do not
   C) nonpolar: do
   D) polar: do
   E) anionic:do

2. If a substance gives off hydroxyl (OH-) ions when dissolved in water, it has a ___ pH and is called a(n) _____.
   A) high : acid
   B) low:acid
   C) high:alcohol
   D) low:base
   E) high:base

3. The charged particles within the nucleus of an atom are called ____.
   A) neutrons
   B) DNA
   C) electrons
   D) glycosides
   E) protons

4. If a bond forms where one atom loses electrons and one gains electrons, the bond is called a ____ bond.
   A) covalent
   B) hydrogen
   C) ionocovalent
   D) ionic
   E) inorganic

5. If two atoms have the same atomic number but different mass numbers, they are described as _____ of the same element.
   A) isomers
   B) orbitals
   C) valences
   D) isotopes
   E) ions

6. Disulfide bonds between cysteine molecules are an example of ___ protein structure.
   A) Primary
   B) Secondary
   C) Tertiary
   D) Quanternary
   E) Duplicative

7. Procaryotic cells lack _____, which eucaryotic cells have.
   A) mitochondria
   B) DNA
   C) cell membranes
   D) ribosomes
   E) mRNA

8. If you discover a new hexose, you have discovered a _____ with six carbons.
   A) protein
   B) lipid
   C) nucleic acid
   D) sugar
   E) amino acid

9. If an atom has 5 electrons in its valence shell, how many chemical bonds can it participate in?
   A) five
   B) one
   C) three
   D) none, its full
   E) four, but only if they are double bonds

10. Proteins can be _____.
    A) enzymes
    B) toxins
    C) antibodies
    D) all of above

11. Which of the following is NOT a type of RNA commonly found in procaryotic cells?
    A) messenger RNA
    B) nRNA
    C) ribosomal RNA
    D) mRNA

12. Which of the following elements is NOT found in every amino acid?
    A) nitrogen
    B) sulfur
    C) oxygen
    D) hydrogen
    E) carbon

13. An anion gains its negative charge by _____.
    A) losing electrons
    B) gaining protons
    C) losing protons
    D) gaining electrons

14. Which of the following is in the correct order, from greatest to least strong bonds?
    A) ionic-covalent-hydrogen
    B) covalent-hydrogen-ionic
    C) hydrogen-covalent-ionic
    D) ionic-hydrogen-covalent
    E) covalent-ionic-hydrogen

15. Phospholipids make up a large portion of cell membranes. They are amphipathic, meaning _____.
    A) they have both proteins and lipids attached to them.
    B) they have two carbon rings, not one carbon ring
    C) they have both hydrophobic and hydrophilic portions

Chemical Connections

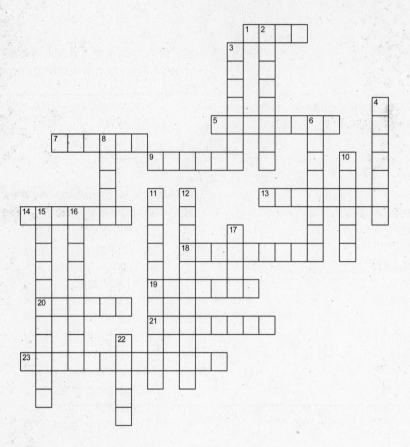

- 14 -

Across

1. The smallest particle of an element that cannot be broken down without losing the properties of the element
5. Strongest bonding type - two atoms share electrons to form a bond
7. Macromolecules made up of fatty acids and glycerol
9. Ions formed when an atom gains electrons
13. Weak bonding interaction between hydrogen and oxygen or nitrogen molecules
14. The outermost shell of electrons in an atom
18. Substances available at the start of a chemical reaction
19. Bond that holds amino acids together to from a protein
20. Nucleotide with a double-ringed nitrogen base
21. Variant forms of an element that have the same atomic number but different mass numbers
23. Lipids that form the semipermeable membranes around cells

Down

2. Level of protein structure represented by disulfide bridges
3. Positively charged subatomic particles in the nucleus of an atom
4. Positively charged ions
6. Neutral subatomic particles in the nucleus of an atom
8. Bonding resulting from the donation and acceptance of electrons
10. Building blocks of carbohydrates
11. Polar substances that dissolve well in water
12. Nonpolar substances that don't dissolve well in water
15. Molecules that have both hydrophilic and hydrophobic properties
16. Negatively charged subatomic particles in orbital shells of an atom
17. A compound that releases hydrogen ions into solution when mixed with water
22. Chemical interactions where two or more atoms share, donate or gain electrons

Chapter 3 Tools of the Laboratory

Building Your Knowledge

1) What are the 5 I's of microbiology? List and describe each.

5 I's	Description

2) How are streak plate, pour plate, and spread plate techniques similar?

Draw the pattern growth you would expect from each technique.

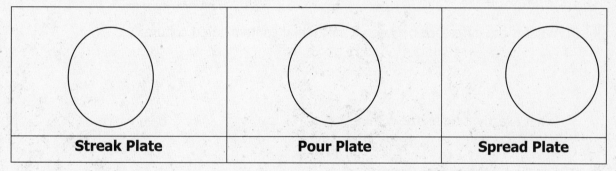

| Streak Plate | Pour Plate | Spread Plate |

3) When is culturing bacteria in broth preferable to culturing on Petri plates?

Can you isolate bacteria in a mixed culture more easily using liquid or solid media?

If you were given a flask with 3 different bacterial species in it, what is the first thing you would do to separate the three?

4) What is agar and why is it so commonly used in microbiology labs?

CHALLENGE – before agar, gelatin was the common medium used to grow isolated colonies, why is agar better than gelatin?

5) How does simple media differ from complex media?

If a media recipe calls for milk or beef extract, are making a synthetic or complex medium?

6) What is the difference between selective and differential media?

If you use a Petri plate with media that changes your target colonies from white to pink, are you using selective or differential media?

7) MacConkey agar is both selective and differential. Explain. It selects for _____ and differentiates between _____ and _____

8) What is the difference between a mixed and contaminated culture?

What exactly is an anexic culture?

9) Labeling – Microscope Parts

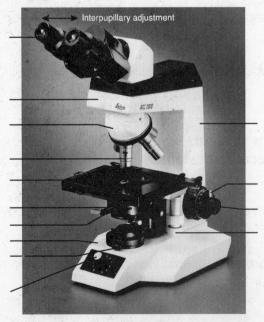

10) How do resolution and magnification differ?

11) Why do bacteriologists use immersion oil when looking at bacteria on microscope slides?

12) If you were a virologist, would you likely be able to see your viral particles with a light microscope? Why or why not?

13) What is confocal scanning optical microscopy, how does it work, and why is of particular use to scientists studying intracellular bacteria, such as *Listeria monocytogenes*?

14) A hanging drop mount is more work to do than a simple wet mount. When would it be better to do a hanging drop mount?

15) Can you Gram stain living cells and look at the living cells under the microscope? Why or why not?

16) What is the difference between positive and negative staining? Would you use acidic or basic dyes for negative staining? Explain your answer.

17) Which staining technique uses a primary dye and counter stain, a simple or differential stains?

18) Describe two types of differential staining. In this description include the primary dye and the counter stain used and what positive and negative samples look like under the microscope.

19) What color are Gram negative bacteria when stained? _____

What color are Gram positive cells? _____

20) Which bacteria are easily identified with an Acid fast stain? (hint – these are the Acid fast bacteria)

21) Name 3 structures that can be identified by staining.

      a. _____

      b. _____

      c. _____

Organizing Your Knowledge

Media Type	Characteristics	Common Use
Liquid		
		Motility
	Media with complex growth factors or nutrients added.	
		Used to study growth with or without oxygen
General Purpose Media		
	Differential media	Types of hemolysis
Transport media		
	Contains sugars to be fermented and a pH indicator	
Liquefiable Solid Media		
Synthetic Media		
		Isolates Gram Negative Enterics
	Media with undefined ingredients, like blood, milk, extracts, infusions.	
Mannitol Salt Agar		

Type of Microscopy	Illuminating Source	Description	Use
	Visible light	Lit specimens surrounded by dark background	
			Visualizing internal contents of live, unstained cells
Phase-Contrast		changes in density are translated to changes in light intensity	
	UV radiation & dyes that fluoresce when activated		labeling structures or bacteria and visualizing them
Scanning Electron		shows surfaces of metal-coated objects in great detail	

Mount or Stain Technique	Live or Dead Specimens?	Characteristics	Use/Examples
	Live	A drop or two of sample with a coverslip	
Hanging Drop			
	Dead		Shows fine detail of internal cellular structures and viruses
		Negative stain with India ink	
Positive Staining			Gram stains, Acid-fast stains
	Dead		To distinguish spores from vegative cells
		A dye technique that stains the background, but not the cells	
Gram staining			

Please fill in the blanks, using your text and notes as a reference.

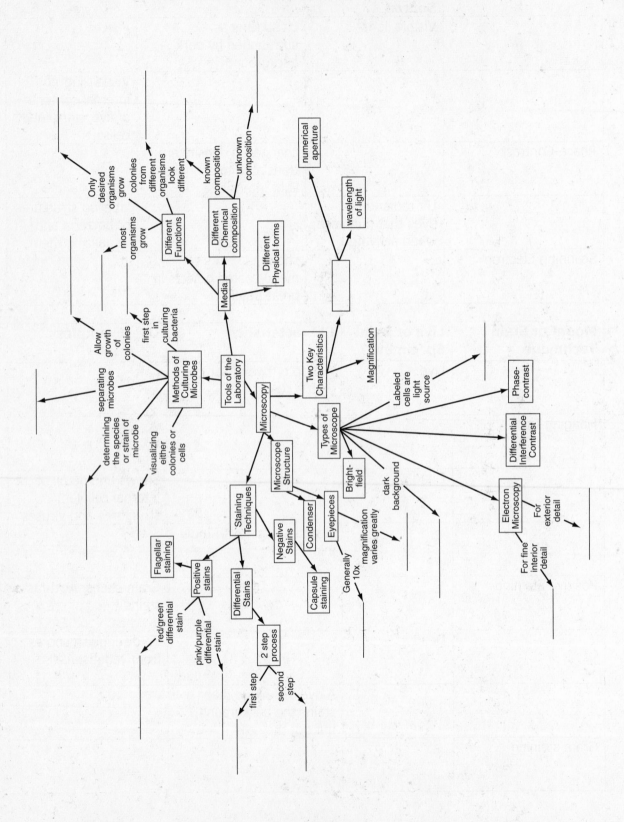

Practicing Your Knowledge

1. Media that grows as many microbes as possible is generally called _____.
   A) selective media
   B) differential media
   C) reducing media
   D) general-purpose media
   E) fastidious media

2. A Petri plate with an unknown fungus growing on it is said to be _____
   A) a mixed culture
   B) an axenic culture
   C) a contaminated culture
   D) a subculture
   E) a gnotobiotic culture

3. Which of the following is NOT one of the "Five Is" culturing microbes?
   A) Inspection
   B) Incubation
   C) Isolation
   D) Irradiation
   E) Inoculation

4. Which of the following staining techniques is NOT a differential staining process?
   A) Gram staining
   B) endospore staining
   C) Loeffler's blue staining
   D) acid-fast staining
   E) All of the above are differential stains

5. If you are making a liquid medium that calls for the addition of meat extract, how could you describe the finished product?
   A) complex agar
   B) synthetic gelatin
   C) complex broth
   D) complex gelatin
   E) synthetic broth

6. The hanging drop slide preparation is commonly used to determine _____.
   A) the species of bacteria you are looking at.
   B) the growth rate of the bacteria you are looking at.
   C) whether your bacteria are Gram negative or Gram positive.
   D) the mobility of the bacteria you are looking at
   E) which antibiotics your bacteria are sensitive to.

7. A differential stain uses _____.
   A) a single basic dye to visualize bacterial cells.
   B) a single acidic dye to visualize bacterial cells.
   C) green capsular dye to visualize viral cells
   D) a primary dye and counterstain
   E) heavy metal dyes to coat target cells for electron microscopy

8. The clarity with which you can see an image is a measure of the _____ of the microscope.
   A) resolution
   B) illumination index
   C) magnification
   D) virtual aperture
   E) field strength

9. If you wanted to isolate several bacterial species from a mixed culture which culture method would you LEAST likely use?
   A) streak plate
   B) spread plate
   C) broth culture
   D) pour plate
   E) nutrient agar

10. Which of the following structures can you NOT commonly see with a light microscope and staining techniques?
    A) bacterial cells
    B) flagella
    C) capsules
    D) bacterial shape
    E) viral particles

11. Acidic dyes are most commonly used to _____
    A) positively stain bacteria for light microscopy
    B) negatively stain bacteria for light microscopy
    C) positively stain fungus for electron microscopy
    D) negatively stain viruses for light microscopy
    E) positively stain viruses for light microscopy

12. A microscope condenser _____.
    A) magnifies an object
    B) focuses light on an object
    C) creates a real image
    D) creates a virtual image
    E) is not part of a compound microscope

13. MacConkey agar is selective because it -
    A) can differentiate between *Salmonella* and *E. coli*
    B) encourages enteric bacterial growth while killing other bacteria
    C) has blood added so *Neisseria* will grow
    D) encourages lactobacilli to grow and kills other bacteria
    E) can differentiate between lactobacilli and coccobacilli

14. Semi-solid media is useful for_____.
    A) determining growth of anaerobes
    B) testing bacterial mobility
    C) determining optimum growth temperatures
    D) determining optimum growth pH
    E) providing micronutrients to growing cultures

15. Which type of microscopy could you NOT use to see living cells?
    A) Bright field
    B) Dark-field
    C) Phase-Contrast
    D) Confocal scanning optical microscopy
    E) Electron microscopy

Chapter 4 Procaryotic Profiles

Building Your Knowledge

1) What were the first cells on Earth? _____

   Which modern cells do they most closely resemble (Archaea/Bacteria/Eukarya)?

   When did these first cells appear?

2) Label the following diagram of a procaryotic cell. Circle the structures found in ALL bacteria and underline those found in MOST bacteria.

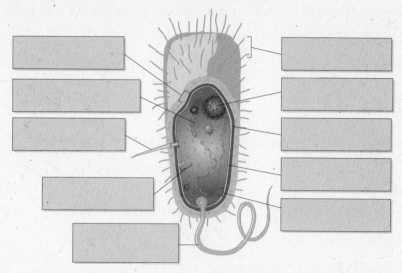

3) Appendages serve two generalized functions for bacteria. What are they?

a. _____

b. _____

4) Draw a bacillus with the following flagellar arrangements.

A) peritrichous	C) monotrichous
B) lophotrichous	D) amphitrichous

Which move faster, bacteria with polar or amphitrichous flagella?

5) How do chemical attractants affect the tumble/run cycle of a motile bacterial cell?

6) In what way are the spirochete flagella unusual? How do spirochetes move?

7) What is the difference between pili and fimbriae? Both are used for _____
   Pili are used for _____ .

8) Draw the three layers of the cell envelop, labeling the interior and exterior of the cell, the glycocalyx, cell wall and cell membrane.

9) Compare and contrast the functions of a slime layer and capsule.

10) What is peptidoglycan and where is it found in bacterial cells?

How are the actions of lysozyme and penicillin similar?

11) Which cells are generally more difficult to kill – Gram negative or Gram positive? Why?

12) What is an Acid-fast stain used for?

Which species of bacteria are Acid fast ?

13) How are L-forms and mycoplasmas similar?

What is the difference between a spheroplast and a protoplast?

14) Why do mycoplasmal cell membranes contain higher levels of sterol molecules?

15) List and describe four separate functions of bacterial cell membranes.

    a. _____

    b. _____

    c. _____

    d. _____

16) There are two structures made of DNA in the bacterial cell.

Which is larger and contains essential genes? _____

Which genes are commonly found on plasmids?

17) What is an endospore?

Why is it an advantage for bacteria to have them?

Name 2 bacteria that form endospores. _____ _____

Are endospores used for reproduction? Explain your answer.

18) List the three most commonly seen bacterial cell shapes.

       a.   _____

       b.   _____

       c.   _____

19) Draw the following arrangements of bacterial cells.

diplococci	streptococci
staphylococci	palisades

20) Compare and contrast microscopic and macroscopic bacterial morphology.

Which process looks at the characteristics of CELLS?

Which looks at colonies or populations of cells?

21) How do tests of metabolic function (biochemical tests) help to identify bacterial samples?

Are these biochemical tests the same as chemical analyses? Explain.

22) Describe three major types of genetic or molecular methods to identify bacteria.

       a.   _____

       b.   _____

       c.   _____

23) Phenetic classification schemes group bacteria based on _____

Phylogenetic classification schemes group bacteria based on _____

24) Bergey's Ninth Edition Manual of Systematic Bacteriology separates bacteria based on

differences in _____. The four major divisions listed are:

    a. _____

    b. _____

    c. _____

    d. _____

25) Are the groupings based on rRNA sequence data the same as those used by Bergey's?

Which groups are similarly arranged?

Which are different?

26) What are the 3 major groups of Archaea, as determined by rRNA sequences?

    a. _____

    b. _____

    c. _____

27) Which system to medical microbiologists commonly use to identify bacteria in clinical

samples? Why?

28) What is a strain of bacteria?

Are two strains of *E.coli* of the same species?

29) Can you grow obligate intracellular parasites on general media agar plate? Why or why

not?

30) Name 2 bacterial obligate intracellular parasites and the diseases they cause.

   a. _____

   b. _____

31) How are cyanobacteria and Green & Purple sulfur bacteria similar?

   Which group produces oxygen in the process?

32) What are myxobacteria and why are they different from most bacteria?

33) Of the three domains, which two are procaryotic?

   a. _____

   b. _____

   Which procaryotic domain is most closely related to eucaryotes?

   Which members of Domain Archaea would you expect to see in the Dead Sea?

Organizing Your Knowledge

Please make an X corresponding to the nature of each trait listed below.

Trait	Microscopic (Requires microscope)	Macroscopic (Naked Eye)
bacterial cell shape		
colony size		
colony shape		
speed of colony growth		
Gram stain		
cell arrangement		
flagellar arrangement		
capsule		
endospores		
slime layers		
colony color		

Procaryotic Profiles

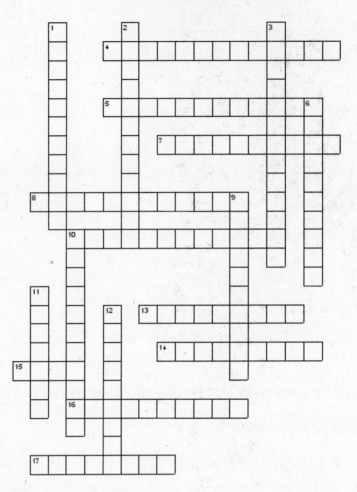

Across

4. sugar-protein complex that is a thick layer in gram-positive bacteria
5. flagella that are randomly dispersed over a cell's surface
7. movement in response to chemical stimuli
8. showing variability in cell shape (e.g. mycoplasmas)
10. strings of circular cells
13. characteristic arrangement of Corynebacteria - resembling a row of fence posts
14. obligate intracellular parasite that causes trachoma, pneumonia, or parrot fever
15. cell appendages that hold bacteria together during conjugation
16. bacterial structures such as flagella and pili
17. classification scheme based on cell morphology or biochemistry

Down

1. heat-loving bacteria
2. structure of a cell consisting of the glycocalyx, cell wall, and cell membrane
3. photosynthesizing bacteria that produce oxygen
6. loose, soluble glycocalyx that is easily removed
9. bacterial species that lack cell walls
10. type of test that uses antibodies to identify bacterial markers called antigens
11. complex association of microbes growing together on the surface of a habitat
12. dormant bacterial structure that is resistant to environmental stress

Practicing Your Knowledge

1. A capsule is used by bacterial cells for all of the following EXCEPT:
   A) conjugation
   B) protection against phagocytes
   C) adhering to surfaces
   D) formation of biofilms

2. Which of the following statements is FALSE, concerning bacterial cell walls?
   A) they have peptidoglycan
   B) they give cells their shape
   C) they protect the cell from hypertonic lysis
   D) they are the target of penicillin action

3. A flagellum is used by a bacterial cell for:
   A) adhesion
   B) structural support
   C) protein synthesis
   D) motility

4. Archaeabacteria include ___.
   A) many human pathogens
   B) mostly flagellated bacteria
   C) extremophiles
   D) all of the Gram negative bacterial species

5. A flagellated bacterial cell moving toward a food source will ___.
   A) make a straight line right for the food.
   B) tumble more than it runs.
   C) run more than tumble
   D) shed its flagella and move with its slime layer

6. If you gram-stain a culture and see purple circles arranged in chains, you would call them:
   A) Gram negative bacilli
   B) Gram positive staphylococci
   C) Gram negative staphylococci
   D) Gram positive streptococci

7. Bacteria are taxonomically classified by___.
   A) cell shape
   B) rRNA sequence similarity
   C) mechanism of mobility
   D) colony morphology

8. Bacterial plasmids will likely carry all of the following genes EXCEPT:
   A) the gene to use a different sugar source
   B) antibiotic resistance genes
   C) genes for the proteins required in metabolism
   D) all of these are commonly seen on plasmids

9. Which of the following structures is NOT found in the cell envelope of a bacterial cell?
   A) cell wall
   B) ribosomes
   C) capsule
   D) glycocalyx

10. If a bacterial cell lost its ribosomes, it would no longer be able to ___.
   A) produce proteins
   B) produce DNA
   C) produce lipids
   D) produce a flagella

11. Which of the following bacteria are photosynthetic?
   A) Cyanobacteria
   B) Chlamydia
   C) Pseudomonas
   D) Treponema

12. Gram positive cell walls ___.
   A) contain LPS
   B) have a thick layer of peptidoglycan
   C) have porins
   D) have an outer membrane

13. Smooth, encapsulated bacteria are generally less pathogenic than are rough bacterial strains.
    A) True
    B) False

14. Which group of bacteria have periplasmic flagella?
    A) bacilli
    B) cocci
    C) vibrio
    D) spirochetes

15. Endospores are used by some bacterial species to reproduce.
    A) True
    B) False

Chapter 5 Eucaryotic Cells and Microorganisms

Building Your Knowledge

1) What is intracellular symbiosis and how does it relate to the evolution of a eucaryotic cell?

2) Describe the movement from simple independent, single-celled eucaryotes to complex multicellular eucaryotes. Which came first, specialization of cell functions or formation of colonies? Why?

3) What is the difference between multicellular organism and a colony of specialized cells?

4) Label the structures found in algae (a), fungus (b), protozoa (c). Use fig. 5.2 as a reference.

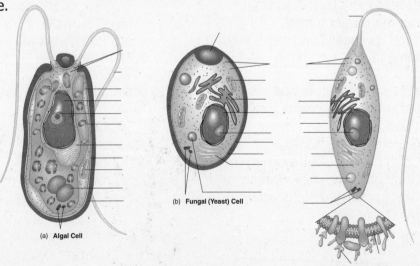

(a) Algal Cell

(b) Fungal (Yeast) Cell

(c) Protozoan Cell

5) Why is mobility important for some organisms?

What 2 structures may propel a protozoan?

_____

_____

6) Are procaryotic and eucaryotic flagella identical in structure and function? Explain your answer.

7) Which are faster, flagellated or ciliated protists? _____

8) What is the function of a glycocalyx?

Where is it found?

9) Fungal cell walls are made up of _____.

What other compounds can make up a cell wall in algae?

Do protozoans have cell walls?

10) Why do eucaryotic cells have many different membrane-bound organelles?

11) One distinguishing characteristic of eucaryotic cells is the presence of a nucleus. Draw a nucleus, labeling the nuclear pores, chromosomes, and nucleolus.

12) Why are chromosomes condensed during mitosis?

What is the purpose of mitosis in eucaryotic cells?

13) If an organism normally has seven chromosomes throughout its life cycle, is this a haploid or diploid species? Explain your answer.

Is this species most likely a sexually reproducing species?

Are gametes (eggs & sperm) haploid or diploid? _____

14) If a cell with 8 chromosomes undergoes mitosis, each resulting cell will have _____ chromosomes.   If that same cell undergoes meiosis, the resulting cells will have _____ chromosomes.

15) What is the difference between rough endoplasmic reticulum and smooth endoplasmic reticulum?

16) Trace the path of a protein for export from the nucleus (where mRNA is formed) to the outside of the cell, labeling the major structures and processes the protein would pass.

17) If a cell lacked lysosomes, what would it not be able to do?

Do you think amoebas and other phagocytes have more or fewer lysosomes than would photosynthetic algae? Explain.

18) Draw a mitochondrion, labeling its inner membrane, outer membrane, matrix, and cristae.

19) What evidence is there that mitochondria and chloroplasts were originally free living procaryotes?

20) How are mitochondria and chloroplasts similar?

What is the function of each?

21) What two cytoskeletal elements are found in eucaryotic cells?

    a. _____

    b. _____

Which moves cilia and flagella?

Which moves pseudopods?

22) What do mycologists study?

23) How have fungus been classified in the past?

How are they classified now?

24) Differentiate between yeasts, hyphae and pseudohyphae.

What are dimorphic fungi?

25) How do fungi acquire nutrition?

Are they autotrophic or heterotrophic?

List 3 substrates fungal enzymes can dissolve.

26) Are most fungi fragile or relatively hardy-growers?

27) If someone or something has a mycosis, what are they infected with?

28) The body of a fungus is called a _____. What is the difference between:

septate and non-septate hyphae?

vegative and reproductive hyphae?

29) Do bacterial endospores and fungal spores serve the same purpose?

30) How are asexual spores produced?

What is the difference between sporangiospores and condida?

31) Are zygospores produced sexually or asexually?

What is the advantage of sexual reproduction?

32) Different fungi have different structures for sexual reproduction. How do zygospores, ascospores and basidospores differ?

Are these spores haploid or diploid?

33) The Mastigomycota live in the _____ and cause disease in _____

and _____.

34) The Amastigomycota are split into four divisions based on _____.

35) What is different about the Deuteromycota, compared to the other three divisions of Amastigomycota?

36) List three types of media that are generally useful in growing fungus.

a. _____

b. _____

c. _____

37) Is spore formation generally used in clinical laboratory identification of fungal cultures? Why or why not?

38) Fungal infections are a particular problem for immunocompromised patients. Why?

39)What are mycorrhizae and are they beneficial or harmful to plants?

40)Which fungus is used in beer brewing and bread making?

41)What is the difference between algae and protozoans?

   Which have cell walls?

42)Where have algae been found growing?

   What are the Five divisions of algae and their common names?

43)Why are dinoflagellates of medical importance to humans?

44)How do protozoans get their nutrients? Are they photosynthetic or heterotrophic?

45)How do protozoans move? What structures may be used?

46)What is the difference between trophozoites and cysts? If a pathogen does not form a
    cyst, would you expect it to be easily transmitted in food or water? Why or why not?

47)Do protozoans reproduce sexually or asexually?

48)Name the four groups of parasitic protozoans and give an example of each. How are
    these divided?

49)How are most clinically relevant human parasites identified in clinical labs?

50)List the 3 stages of most human parasites' life cycles.

        a. _____

        b. _____

        c. _____

51) How are trypanosomes spread?

What diseases do they cause?

52) How is amebic dysentery spread?

What are the major symptoms of the disease?

53) What are helminthes?

Are they free-living or parasitic?

Which structures are most pronounced in helminthes (circulatory, nervous, reproductive, digestive, etc.)?

54) How are helminth infections identified in the laboratory?

Organizing Your Knowledge

Please make an X corresponding terms describing the following organisms.

Organism or Group	Multicellular?	Colonial?	Eucaryotic?	Procaryotic?
Archaea				
Mushrooms				
Animals				
Most Algae				
Plants				
Bacteria				
Seaweeds				
Protozoa				

Fungal Group	Sexual Reproductive Structures	Common Examples
Zygomycota		
Ascomycota		
Basidiomycota		
Deuteromycota		

Group	Organization	Cell Walls made of?	Examples/Importance
Euglenophyta			
Pyrrophyta			
Chrysophyta			
Phaeophyta			
Rhodophyta			
Chlorophyta			

Structure	Found in: Eucaryotes or Procaryotes or Both	Function?
smooth endoplasmic reticulum		
flagella		
nucleus		
rough endoplasmic reticulum		
glycocalyx		
Golgi apparatus		
cilia		
mitochondria		
cell wall		
plasma membrane		
nucleolus		
microtubules		
chloroplasts		
ribosomes		

Practicing Your Knowledge

1. Red tides are caused by overgrowths of ___.
   a. red algae
   b. diatoms
   c. dinoflagellates
   d. apicomplexans

2. A medical mycologist studies:
   a. mycorrhizae on plant roots
   b. fungus than infect humans
   c. algae that infect fish
   d. yeasts that are used to brew beer

3. Most water-transmitted protozoan parasites have a well-developed ____ stage.
   a. trophozoite
   b. cyst
   c. endospore
   d. sarcodinan

4. If a cell has 16 chromosomes at the start of mitosis, how many will be in each cell at the end of mitosis?
   a. 8
   b. 16
   c. 24
   d. 32

5. The cell walls of fungus are made up of:
   a. chitin
   b. cellulose
   c. silicon dioxide
   d. cellulose

6. Which eucaryotic organelles were likely originally free-living bacteria?
   a. chloroplasts & mitochondria
   b. nucleus & lysosomes
   c. Golgi apparatus & nucleus
   d. nucleolus & flagella

7. Protozoan may move by all of the following EXCEPT_____.
   a. pseudopods
   b. flagella
   c. pili
   d. cilia

8. If an organism has 5 chromosomes most of the time, we can say its ___.
   a. a diploid organism
   b. a bacterium
   c. a haploid organism
   d. has cilia, not flagella

9. Trypanosomes are important protozoan parasites transmitted by ____ and cause ____ .
   a. mosquito: malaria
   b. water: amebic dysentery
   c. insect bite: sleeping sickness
   d. feces :tapeworm infection

10. The nucleolus is the ___.
   a. area where lysosomes are made
   b. site of ribosome synthesis
   c. central part of a procaryotic cell
   d. site of flagellar attachment

11. The Amastigomycota, or terrestrial fungi are organized based on their ____.
   a. size and color
   b. asexual reproductive mode
   c. digestive mode
   d. sexual reproductive mechanism

12. Which of the following pairs is mis-matched?
   a. flagella - movement
   b. lysosomes - protein degradation
   c. cilia - adhesion
   d. ribosomes- protein synthesis

13. Which of the following are NOT eucarytotic?
   a. fungus
   b. algae
   c. protozoa
   d. bacteria

14. Which of the following organelles does a protein NOT go through on its way out of a cell?

    a. lysosomes
    b. rough ER
    c. vesicles
    d. Golgi apparatus

15. Most parasites dedicate a great deal of body-space to _____ structures.

    a. reproductive
    b. digestive
    c. circulatory
    d. nervous system

Chapter 6 An Introduction to Viruses

Building Your Knowledge

1) Can you see viral particles with a light microscope?

2) What does the term virus mean in Latin? Who coined the term virus?

3) When were viruses first discovered as plant and animal pathogens?

    Which virus was isolated?

    How were viruses separated from bacteria?

4) What cell types can be infected with viruses?

5) Are viruses alive? Explain your answer.

    Why are viruses called "particles" and not cells?

6) All viruses are obligate intracellular parasites. What does mean?

    Can viruses be grown on nutrient agar? Why or Why not?

7) How large are the smallest bacterial cells, such as rickettsias?

    How large are the largest viruses?

    Are rickettsias free-living or parasites? Are viruses free-living?

8) Draw two viral particles, labeling the capsid, nucleic acid, envelope, and spikes.

naked	enveloped

9) A viral particle is made up of _____ and _____.

10) What are capsomers and where are they found on a viral particle?

11) Do all viruses have an envelope?

What is a viral envelope made up of (DNA, sugar, lipids, protein)?

Where does the viral envelope come from?

12) Are viral spikes found on naked viruses?

What are these spikes used for?

13) What 3 functions do capsids and envelopes perform for a viral particle?

a. _____

b. _____

c. _____

14) What else may be found inside a viral particle, with capsid proteins, a nucleic acid core, and an envelope?

15) Are DNA and RNA ever found in the same viral particle?

16) What are the three major criteria used to classify viruses?

a. _____

b. _____

c. _____

17) Differentiate between a lytic and lysogenic viral cycle.

Which kills the host cell when it exits?

Which does not?

18) Label the figure below with these terms: adsorption, penetration, assembly, maturation, release.

19) If you lyse cells after penetration, but before maturation, would there be active viral particles released? Why or why not?

What do we call this stage in the viral cycle?

20) Is adsorption of phage to an animal cell a specific or generalized process?

How does this affect viral host range?

21) What happens to phage DNA if the phage is a temperate (or lysogenic) phage?

22) What is transduction and how do phage play a role in the process?

23) Why does hepatitis B only infect human liver cells?

24) Give examples of human only and of multiple-host viruses.

   human only -

   multiple-host -

25) How does a viral particle enter a human host cell?

26) Where are DNA viruses assembled in animal cells?

   Where are RNA viruses assembled in animal cells?

27) Do enveloped viruses lyse animal cells as they exit?

   Why or Why not?

   Do naked viruses lyse animal cells?

28) Name two persistent viral infections and the diseases they cause.

   a)_____

   b) _____

29) How can viral infection lead to cancer formation? Name three specific viruses that have been shown to cause cancer.

30) Why do scientists cultivate viruses? List three reasons.

31) How are viruses cultured? List three separate ways.

      a. _____

      b. _____

      c. _____

32) List five different diseases caused by viruses.

      a. _____

      b. _____

      c. _____

      d. _____

      e. _____

33) What are prions?

What diseases do they cause?

How do they differ from viruses?

34) What are viriods?

What diseases do they cause?

How do they differ from viruses and prions?

35) Why are antibiotics ineffective against viral infections?

36) Most anti-viral drugs are toxic to the host as well as the virus. Why do you think this is the case?

Practicing Your Knowledge

1. Where does an enveloped virus's envelope originate?

    a. the host cell's plasma membrane
    b. the capsid of a second virus
    c. the nuclear envelope of bacterial cells
    d. the inner membrane of host cell mitochondria

2. Viruses are commonly visualized using a light microscope on its highest setting.

    a. True
    b. False

3. Transduction is a form of gene transfer in bacteria.

    a. True
    b. False

4. Spikes are found on _____ viruses and are used for _____.

    a. naked viruses: movement
    b. enveloped viruses : adhesion
    c. naked viruses: adhesion
    d. enveloped viruses: movement

5. If you lyse cells during the eclipse period, _____

    a. many active virus particles will be released
    b. enveloped viruses will become infectious naked viruses
    c. naked viruses will become dormant enveloped viruses
    d. no infective virus particles will be released

6. Which of the following is NOT used to cultivate viruses?

    a. tissue culture cells
    b. eggs
    c. blood agar
    d. mice

7. Lysogenic phage ____.

    a. lyse the host cell immediately after entry
    b. infect cells and integrate their DNA into the host cell's DNA
    c. do not exist in nature
    d. all of the above are correct

8. Which of the following can form a latent viral infection?

    a. herpes virus
    b. common cold
    c. lytic bacteriophage
    d. there are no latent infections in nature

9. A viral particle is made up of ____.

    a. DNA and RNA in the same particle
    b. DNA & lipids
    c. DNA or RNA and proteins
    d. sugars and lipids

10. Tissue tropism determines _____.

    a. which tissues a virus will infect
    b. how long a virus will take to kill a cell
    c. whether a particle is a DNA or RNA virus
    d. which tissues will die first

11. Hepatitis B only infects human liver cells because ____.

    a. only liver cells have enough DNA to infect
    b. only human cells have plasma membranes
    c. only liver cells have the right receptors on their surface
    d. only human cells have a nucleus

12. Where do DNA viruses that infect animal cells mature?

    a. in the nucleus
    b. under the cell wall
    c. in the cytosol
    d. in the chloroplasts

13. Which of the following organisms can serve as a host for a virus?

    a. bacteria
    b. fungus
    c. animal
    d. all of the above

14. Prions are ____.

    a. infectious DNA viruses
    b. dormant RNA viruses
    c. infectious proteins
    d. infectious RNA

15. Enveloped viruses leave animal cells ____.

    a. by disrupting the cell
    b. by budding off the plasma membrane
    c. by disrupting the nucleus which kills the cell
    d. enveloped viruses never leave animal cells

Chapter 7 Elements of Microbial Nutrition, Ecology and Growth

Building Your Knowledge

1) What is the process by which organisms acquire nutrients from the environment?

2) How do essential nutrients differ from non-essential nutrients?

3) Differentiate between micronutrients and macronutrients.

    Which are needed in greater quantity?

    _____

    What is the function of each in a bacterial cell?

Micronutrients _____

Macronutrients _____

4) What is the difference between heterotrophs and autotrophs?

Autotrophs get carbon from _____

Heterotrophs get carbon from _____

5) How do photoautotrophs and chemoautotrophs differ in how they get energy?

Photoautotrophs get energy _____

Chemoautotrophs get energy _____

6) Photosynthetic algae get carbon and energy from _____.

7) Humans and most bacteria are _____ (autotrophs or heterotrophs).

8) What is the difference between saprobes and parasites/pathogens?

9) Can saprobes cause disease?

10) Can obligate parasites be cultured using solid, synthetic media? Explain your answer.

11) Name an obligate parasite.

12) All viruses and some bacteria are obligate intracellular parasites. Name two bacteria that are obligate intracellular parasites.

    a. _____

    b. _____

13) Without the addition of energy, do molecules move from high concentrations to low concentrations or from low concentrations to high concentrations? Explain your answer.

14) Draw a cell in hypertonic, isotonic and hypotonic solutions. Use x's to indicate solute
molecules. Which cells shrink? Which swell?

Hypertonic	**Isotonic**	**Hypotonic**

What does a bacterial cell wall protect against?

15) How are facilitated diffusion and active transport similar?

How are they different?

Do both require energy?

16) When is active transport necessary?

What advantage does group translocation have over simple pumps?

17) What are the three cardinal temperatures of microbial growth?

a. _____

b. _____

c. _____

18) Do all microbes have similar temperature ranges? Explain.

19) What are psychrophiles, mesophiles, and thermophiles?

Which are of concern to food microbiologists?

Which are most commonly pathogenic?

20) How do the oxygen requirements of obligate aerobes, facultative anaerobes, microaerophiles and obligate anaerobes differ?

21) How would each grow in thioglycolate broth Draw each test tube.

Obligate aerobe	Microaerophile	Obligate anaerobe	Facultative Anaerobe

22) Halophiles live in extreme _____ conditions.

23) Barophiles live in extreme _____ conditions.

24) How do bacteria reproduce? Draw the process of binary fission below.

25) Do all bacteria reproduce at the same rate? Explain and give examples.

26) Draw a growth curve, labeling lag phase, log phase, stationary phase and death phase.

Time

27) Which stage has the fastest growing bacteria?

28) Why do cultures move from log phase to stationary phase?

29) Why do cultures move from stationary to death phase?

30) If you place 100 bacterial cells in media and their doubling time is 30 minutes, how many cells are in the media at the end of 5 hours?

31) How may we count bacteria? (list 3 ways) Which methods count live cells only?

    a. _____

    b. _____

    c. _____

Live cells only - _____

Organizing Your Knowledge

Type of Bacteria	Living Conditions preferred
psychrophile	
	Acidic pH
obligate anaerobe	
	Small amounts of oxygen
Alkalinophile	
	Extreme salt conditions
Osmophile	
	Moderate temperatures
Barophile	
	Extreme Heat
Obligate Aerobe	
	Can grow with or without oxygen

Micronutrient	Use in bacterial cells
Potassium	
Sodium	
Calcium	
Magnesium	
Iron	

Macronutrient	Use in Bacterial Cells	Source or Environmental reservoir
Carbon		Autotrophs
		Heterotrophs
Nitrogen		
Oxygen		
Phosphorus		
Sulfur		

Relationship	Interaction (+/+, +/-, +/0, etc.)	Symbiotic (yes/no?)
mutualism		
commensalism		
parasitism		
synergism		
antagonism		

Stage of Growth Curve	What's Happening?	Growth Speed (fast/slow/level)	Live/dead cell ratio
Lag phase			
Exponential phase			
Stationary phase			
Death phase			

Practicing Your Knowledge

1. If placed in a hypertonic solution, most bacterial cells will_____.

    a. burst if it lacks a cell wall
    b. remain unchanged
    c. shrink and die
    d. change color

2. Bacteria preferring low temperatures for optimum growth are called_____.

    a. barophiles
    b. halophiles
    c. thermophiles
    d. psychrophiles

3. Macronutrients are required by cells in ___ quantities and are used to ___.

    a. small: boost enzyme function
    b. large: boost enzyme function
    c. small: form cell structures
    d. large: form cell structures

4. Which of the following methods measures live bacterial cells only?

    a. turbidity
    b. plate counts
    c. cytometer
    d. Coulter counter

5. Which of the following transport processes requires energy?

    a. diffusion
    b. osmosis
    c. facilitated diffusion
    d. group translocation

6. Which of the following microbial associations is NOT symbiotic?

    a. mutualism
    b. commensalism
    c. synergism
    d. parasitism

7. Which phase of the growth curve sees an equal rate of bacterial death and reproduction?

    a. lag phase
    b. stationary phase
    c. exponential phase
    d. death phase

8. Strict halophiles are commonly human pathogens.

    a. True
    b. False

9. Photosynthetic bacteria are considered ___.

    a. non-existent - bacteria don't have chloroplasts
    b. heterotrophs because they feed off dead things
    c. autotrophs because they get their carbon from carbon dioxide
    d. saprobes because they feed off dead things

10. As a bacterial culture grows, the media _____.

    a. gets thicker because of all bacteria
    b. gets cloudier because of all the bacteria
    c. gets warmer because of the heat generated by bacterial cells
    d. gets clearer because the bacteria consume all the nutrients

11. What is the correct order for a growth curve progression, with bacterial cells in batch culture?

   a. lag phase - exponential - stationary
   b. stationary - lag phase - exponential
   c. exponential - stationary - lag phase
   d. lag phase - stationary - exponential

12. Bacteria lacking superoxide dismutase and catalase are ___.

   a. strict aerobes
   b. strict anaerobes
   c. facultative anaerobes
   d. strict acidophiles

13. What are the three cardinal temperatures for microbial growth?

   a. hypertonic, isotonic and hypotonic
   b. minimum, maximum and optimum
   c. aerobic, anaerobic and microaerobic
   d. halophile, barophile, osmophile

14. Active transport mechanisms are required to ___

   a. move nutrients from high concentrations to low
   b. move any nutrient across a plasma membrane
   c. complete facilitated diffusion
   d. move molecules from low concentrations to high

15. Most human pathogens are _____.

   a. mesophiles
   b. psychrophiles
   c. thermophiles
   d. psychrotrophs

Chapter 8 Microbial Metabolism

Building Your Knowledge

1) What are the two branches of metabolism?

   a. _____

   b. _____

   Which branch synthesizes large molecules from small subunits? _____

   Which breaks down large molecules into small subunits? _____

2) What do catalysts do in a chemical reaction?

3) Do enzymes add energy to chemical reactions?

Are they changed by the reaction?

Do they interact with several substrate molecules or one molecule per enzyme (then the enzyme goes away)?

4) What are enzymes made up of –proteins, lipids, or sugars? _____

5) How do enzymes speed chemical reactions?

6) How do enzyme co-factors and co-enzymes differ?

How are they similar?

7) Endoenzymes work inside the cell. What are enzymes that work outside a cell called?_____

8) Enzymes that are present all the time are called _____.

Induced enzymes are activated or produced only when _____ is present.

9) The removal of water is a _____ reaction.

10) The addition of water _____ chemical bonds.

11) How are oxidation and reduction related?

If a molecule is reduced, does it gain or lose electrons? _____

If a molecule is oxidized, does it gain or lose electrons? _____

12) How do enzymes contribute to the disease process caused by *Streptococcus pyogenes*, *Pseudomonas aeruginosa*, and *Clostriduium perfringes*?

Organism	Enzyme	Disease
Streptococcus pyogenes,		
Pseudomonas aeruginosa		
Clostriduium perfringes		

13) What is energy?

    List 3 forms of energy.

        a. _____

        b. _____

        c. _____

    Which forms of energy are most commonly used in cells?

14) How do endergonic and exergonic reactions differ?

    Which are typically anabolic? _____

    Which are typically catabolic?_____

15) What is ATP and why is it called "metabolic money"?

    Label the following figure, indicating the location of the 3 phosphate groups, ribose and nitrogen base (adenine)

16) Which yields more energy, anaerobic respiration or aerobic respiration?

    Which requires oxygen?

17) What is the basic equation for aerobic respiration in microbes?

    For every glucose molecule burned, the cell needs _____ oxygen molecules, and produces _____ molecules of carbon dioxide and _____ molecules of water.

18) What is the final electron acceptor in aerobic respiration?_____

19)Does the TCA cycle reduce or oxidize glucose? _____

20)Glycolysis starts with _____ and ends with _____.

How many ATP molecules are generated in glycolysis for each molecule of glucose consumed?

21)How many carbons are in a glucose molecule? _____

How many carbons are in a pyruvic acid molecule? _____

How many pyruvic acids are produced for every glucose molecule metabolized?_____

22)The TCA cycle produces _____ and_____.

Where do NADH molecules go with their electrons?

23)Which stage of glucose metabolism requires a membrane?

Why?

24)How does ATP synthase generate ATP?

25)Draw an ATP synthase molecule, a membrane, the H+ gradient, the flow of H+ ions and the formation of ATP from ADP and P.

26) Why do we consider pyruvic acid a central part of metabolism?

What can pyruvate be converted to anaerobically?

Do all bacteria use pyruvate in the same way?

27) Which form of glucose metabolism yields more energy – anaerobic or aerobic?

Where is most ATP generated?

28) How do fermentation and anaerobic respiration differ?

Which yields more energy per glucose molecule?_____

29) How do alcoholic and acidic fermentation differ?

Which fermentation do you want if you are making bread or beer?
Which process sours milk?

Which process do you want if you are making yogurt?

What happens when you work out to the point your muscles are deprived of oxygen?

30) Amino acids are made up of carbon and nitrogen. Where can cells get the carbon?

What do amino acids combine to form?

31) How are carbohydrates produced?

Where are carbohydrates used in a bacterial cell?

32) How are lipids (fats) made?

What are they used for in a procaryotic cell?

33)Many metabolic pathways are amphibolic. What does this mean?

34)Do pre-cursor molecules (amino acids, sugars, lipids) come from the electron transport chain (Yes or No)?

Where may they come from?

35)If we labeled a glucose molecule's carbon atoms radioactively, so they could be traced, where would the carbons exit the metabolic pathway?

Organizing Your Knowledge

Part of Aerobic Respiration	Location	Starting molecules	End products
glycolysis			
TCA cycle			
Electron transport chain			

Metabolic Mechanism	Pathways Included	Final Electron Acceptor	Products	Microbes Using this
Aerobic respiration				
Anaerobic fermentation				
Anaerobic respiration				

Practicing Your Knowledge

1. Enzymes _____.

    a. add energy to chemical reactions
    b. increase the rate of chemical reactions
    c. are changed by the chemical reactions they catalyze
    d. work on all chemical reactions the same way

2. What is the final electron acceptor in aerobic respiration?

    a. oxygen
    b. carbon dioxide
    c. sulfur
    d. NADH

3. Which of the following factors will change enzyme function?

    a. temperature
    b. pH
    c. substrate concentration
    d. all of the above

4. An enzyme inhibitor that binds to the site normally used by a substrate and blocks enzyme function is called a _____.

    a. positive feedback inhibitor
    b. competitive inhibitor
    c. allosteric inhibitor
    d. enzyme inducer

5. The energy of activation of a chemical reaction _____

    a. increases when enzymes are present
    b. decreases when enzymes are present
    c. is not changed by enzymes

6. Beta-galactosidase is not produced by a cell unless its substrate, lactose is present. It metabolizes lactose inside the cell. We would describe this as a _____ enzyme.

    a. constitutive endoenzyme
    b. induced endoenzyme
    c. induced exoenzyme
    d. constitutive exoenzyme

7. Enzyme cofactors are _____.

    a. generally vitamins and used to support enzyme function
    b. generally apoenzymes and work alone
    c. generally metallic and activate enzymes
    d. not used in bacterial cells, procaryotes have simple enzymes

8. If you labeled the carbons of glucose and sent it through aerobic respiration, where and how would the carbons be released?

    a. in glycolysis as carbon dioxide
    b. in glycolysis as water
    c. in the TCA cycle as water
    d. in the TCA cycle as carbon dioxide

9. Which portion of aerobic respiration requires a membrane to generate energy?

    a. glycolysis
    b. TCA cycle
    c. electron transport chain
    d. fermentation

10. Which part of central metabolism does NOT contribute precursor molecules to anabolic pathways?

    a. TCA cycle
    b. electron transport chain
    c. glycolysis
    d. pyruvic acid

11. The loss of electrons is a(n) ___

    a. reduction
    b. oxidization
    c. condensation
    d. induction

12. The addition of water to chemical bonds ____ them and is a ____ reaction.

    a. creates : anabolic
    b. breaks: anabolic
    c. creates:catabolic
    d. breaks:catabolic

13. Anabolic reactions ____ energy and are used in a cell for ____ reactions.

    a. release: synthesis
    b. use:degradative
    c. release: degradative
    d. use: synthesis

14. ___ is the energy currency of cellular reactions.

    a. DNA
    b. phosphate
    c. ATP
    d. AMP

15. Where is most of the energy (ATP) generated during aerobic respiration?

    a. glycolysis
    b. TCA cycle
    c. fermentation
    d. electron transport chain

Microbial Metabolism

```
P W K U E J Y C W D Y Z T Z F Q X A M E Q R O K R R G R N Q
L D O K G Y Y X T P S I G I T Y T G V I A O M M B B N K A A
R Q Z U M B B W L K P I T N C W C Q Y S A N X F Q W N A B I
C Z N B N E F Z W Q J Z L G Z S C Y M X R K X K U J D Q L
Z O Y S C R V S L G D Y P T R D U Z K G E U K W E O Y L H X
P W H Z D C W G L Y C O L Y S I S B X C C T F H I S B E B F
M O L E N A C O N S T I T U T I V E S M X N A P K Q Q D O H
R T K J O H M F X F W A L O Z K H L U T A U P B E H K W U L
V L D W C G T D R K C B C U Y F E O Z E R U E O O R A H T E
B K H Q T G E J E I V P A T C K N D M E R A D N X L Y N A C
F X X R M Y F P W R I D T D X X J A D F W E T R J Y I Z Z H
A Q Z K J K P W K D A M D J E W G Y Q K X D E F N T S U Z
H J E P Q M E C O M P E T I T I V E I N H I B I T I O N M W
Z Y W O U H N T I J H B A M E V E M K L F I Z G N E P K E A
Z O D P O E E A B C T E V J O Z S V N W E S N R C A P F V W
B V Z R Q K R B G L A F C E B V B S K S R Q C D A G C C W A
M Q H I O Z G W T W U K Q F K F S T T F M O S K U Y N Q Q K
U L U S V L Y L F V Y D B D V Y O R T I E O J D I C X R O D
J C P G H N Y O F Z P J E D I S E N T X N C Q O Z E E W H X
Y S B N T R F S W E J B J V S P G X J F T B H P C P C D R O
C U N R C P W H I V U W T G T Q A F Y K A B A J Q A D T N W
A N A B O L I S M S Y L P O L U M D W X T R R A T O I W K F
L Y T C A C Y C L E H J K O L E Y R V B I J X A E N Z Y M E
G A D W X Q H M F Y P I D F P E Z J V H O Q B A F I L C V D
G V B X L N X I S W N W Y T D R F L C Z N O V V S U K H X N
U J G I W O C C L A A M P H I B O L I C L C O F A C T O R S
V X A Z L D H H S R E D U C T I O N F I R S E N V O E B A S
S I O H L E E E W N O M V Y E G W D S Q O E N O E K O T F T
S N X E N D E R G O N I C Z T O O M P L L W Y F O G P F L O
O O U L O S N Z Q B Q V P I V F L F T O Z J K M G J L P G S
```

1. (_ _ _) Energy currency of a cell
2. (_ _ _ _ _ _ _ _) Cycle that takes in pyruvic acid and converts it to CO2 and provides NADH for the electron transport chain
3. (_ _ _ _ _ _ _ _ _ _) Metabolic pathways that can be used for anabolism and catabolism
4. (_ _ _ _ _ _ _ . _ _) Synthesis of large molecules from small ones
5. (_ _ _ _ _ _ _ _ _) Breakdown of large molecules into small ones
6. (_ _ _ _ _ _ _ _ _) Metallic ions associated with enzymes that are critical to enzyme function
7. (_ _ _ _ _ _ _ _ _ _ _ _ _ _ _ _ _ _ _ _) Process by which a substance binds to the active site of an enzyme and stops it from binding to its substrate
8. (_ _ _ _ _ _ _ _ _ _ _ _) Enzyme that is present in constant concentrations, independent of substrate concentration
9. (_ _ _ _ _ _ _ _ _ _) Chemical reaction that requires the addition of energy
10. (_ _ _ _ _ _) The ability to do work
11. (_ _ _ _ _ _) Protein catalyst that speeds reactions by lowering the energy of activation
12. (_ _ _ _ _ _ _ _ _ _ _ _) Anaerobic process that leads to the production of gases, acids and/or alcohol
13. (_ _ _ _ _ _ _ _ _ _) Pathway that converts glucose to pyruvic acid
14. (_ _ _ _ _ _ _ _ _ _) The addition of water to break bonds
15. (_ _ _ _ _ _ _) Enzyme that is found in higher concentrations when its substrate is present
16. (_ _ _ _ _ _) Molecules that are chemically unstable are called _____
17. (_ _ _ _ _ _ _ _ _ _) Sum of all chemical and physical activities - converting energy to usable forms AND using energy to do work
18. (_ _ _ _ _ _ _ _ _) The gaining of electrons
19. (_ _ _ _ _ _ _ _ _ _ _ _ _) Product of Streptococcus pyogenes that dissolves blood clots
20. (_ _ _ _ _ _ _ _ _) Molecule that enzymes interact with to produce products

- 61 -

Chapter 9 Microbial Genetics

Building Your Knowledge

1) The study of genetics explores many different sub-topics. List four of these sub-topics that geneticists study.

   a. _____

   b. _____

   c. _____

   d. _____

2) What is a genome?

   How may genomes be arranged?

3) What is a gene?

   How many genes and chromosomes are in an *E.coli* genome?

   How many genes and chromosomes are in the human genome?

4) Is DNA longer or shorter than the cells that hold it?

5) What the 3 basic components of DNA?

   Draw a nucleotide, labeling the 3 parts. (deoxyribose, nitrogenous base, phosphate). Number the carbons on the dexoyribose 1-5.

6) What are complementary bases?

   Do purines bind with each other or with pyrimidines?

   What type of bonds hold complementary bases together?

   What are the complementary base pairs in DNA? A bonds with _____ and C binds with _____.   Are they the same in RNA?

7) What bonds hold the DNA backbone together?

Are these stronger or weaker than the bonds that hold the complementary bases together?

8) Draw a double-stranded DNA molecule, 6 nucleotides long, with complementary bases and number of hydrogen bonds for each indicated and the orientation of each strand (3' and 5').

9) What does it mean that the strands of DNA are anti-parallel to one another?

10) Why is it important that the nitrogenous bases in DNA have complementary pairs and that there are 2 different purines and 2 different pyrimidines? (give 2 reasons)
   a.

   b.

11) What enzymes are required for DNA replication and what is the function of each?

Enzyme Required	Function

12) If a bacterial cell was deficient in DNA polymerase I, would you expect greater or fewer mutations? _____
Explain your answer.

13) Draw the replication bubble of bacterial DNA replication, include in your drawing the origin of replication and all enzymes and binding proteins required for replication.

14) Does RNA polymerase require a 3' OH to produce RNA from DNA? _____

Why is this important to DNA synthesis?

15) Why is DNA synthesized with a leading and a lagging strand?

How many complete strands would be formed at the end of DNA replication if a cell lacked ligase activity?

16) What is the rolling circle model of DNA synthesis? Give an example of genetic material that uses this model of synthesis.

17) What are the 3 basic categories of genes?

a. _____

b. _____

c. _____

18) How is RNA different from DNA? Give 3 specific reasons.

a. _____

b. _____

c. _____

19) What are the three stages of transcription?

    a. _____

    b. _____

    c. _____

What happens during each?

20) What are the 3 types of RNA found in a cell, what is the function of each and are these translated?

RNA	Function	Translated?

21) Draw the major steps in translation. Label the coding strand, template strand of DNA, RNA polymerase, direction of transcription and the growing mRNA transcript.

22) What are the 3 stages of translation?

    a. _____

    b. _____

    c. _____

23) What is a codon?

24) Why is the genetic code of mRNA codons considered universal?

25) Draw a ribosome, with an mRNA molecule and its A and P sites filled.

26) Given the RNA sequence AUG UUA CUA CCG GCG UAG what would the amino acid sequence look like?

27) What is a promoter?

28) What happens when a ribosome encounters a nonsense codon on an mRNA message?

29) What is a polyribosome?

Do polyribosomes exist in eucaryotes?

30) What are introns and exons?

Do bacterial cells have introns and exons?

31) How long does it take an average protein to be translated?

How much energy (in ATP expenditure) does it take to make an averaged protein?

32) Where are double stranded DNA (dsDNA) viruses replicated in animal cells?

How are dsDNA viruses replicated (where does the polymerase come from?)

What may integration of viral DNA do to the host cell?

33) How do positive sense ssRNA viruses and negative sense ssRNA viruses differ?

Which RNA viruses are ready to be translated when they enter a host cell?

Which RNA viruses use reverse transcriptase?

34) Diagram the lac operon.
   a. What does it look like when lactose is not available?

   b. What does it look like when lactose is available?

35) How may antibiotics affect transcription and translation in prokaryotes?

36) What is replica plating used for? Why do you use selective media to look for mutants?

   What is the Ames test used for?

   What are the steps in the Ames test?

37) Diagram the effects of deletions, insertions, inversions and duplications on a stretch of
   DNA coding for 3 genes, A B C.
   Original   A   B   C

   Insertion _____

   Inversion _____

   Deletion _____

   Duplication _____

38) Compare and contrast same-sense, mis-sense and nonsense mutations.

39) Frame-shift and non-sense mutations often knockout expression of a particular gene. Why do you think this is the case?

40) What is genetic recombination in bacteria?

41) How may genes be transferred from one bacterial cell to another? List and define 3 processes.

Transfer process	Definition

42) Diagram the Griffith Experiment on bacterial transformation and discuss what it showed.

43) Differentiate between generalized and specialized transduction.

Generalized -

Specialized - ?

Organizing Your Knowledge

Nucleotide	Purine or Pyrimidine?	Pairs With	Purine or Pyrimidine
thymidine			
guanine			

Enzyme, protein or factor	Purpose	Process its involved in
helicase		
RNA primer		
primase		
DNA polymerase I		
DNA polymerase III		
ligase		
gyrase		
RNA polymerase		
codons		
polyribosomal complex		
operon		
mRNA		
rRNA		
tRNA		

Template	Product	Enzyme	Process
DNA	RNA		
RNA	DNA		
mRNA	protein		
DNA	DNA		

Practicing Your Knowledge

1. Replication is _____.

    a. The production of DNA from a DNA template.
    b. The production of RNA from a DNA template
    c. The production of RNA from a RNA template
    d. The production of DNA from a RNA template

2. Okazaki fragments are ___.

    a. the normal result of leading strand synthesis
    b. the result of mistakes in transcription
    c. the normal result of lagging strand synthesis
    d. the result of mistakes in translation

3. The sum total of the genetic material of a cell is called its _____.

    a. plamids
    b. chromosome
    c. polyribosomal complex
    d. genome

4. The addition of 1 or 2 bases in a DNA message will often lead to a _____.

    a. translocation
    b. inversion
    c. frame-shift
    d. compensatory deletion

5. The complementary base pairs of DNA are held together by___.

    a. covalent bonds
    b. hydrogen bonds
    c. ionic bonds
    d. ionocovalent bonds

6. RNA polymerase binds to ____.

    a. The DNA of the promoter region
    b. The RNA of the promoter region
    c. The DNA of the start codon
    d. The RNA of the start codon

7. The formation of a polyribosomal complex is _____.

    a. part of normal DNA replication
    b. part of abnormal RNA transcription
    c. a nonsense mutation
    d. a part of normal translation

8. Which of the following is NOT a form of RNA found in cells?

    a. ribosomal RNA
    b. transfer RNA
    c. messenger RNA
    d. lysosomal RNA

9. Which of the following nucleotides are purines?

    a. adenine and thymine
    b. cytosine and guanine
    c. adenine and guanine
    d. cytosine and adenine

10. Which method of gene transfer between procaryotes requires DIRECT contact?

    a. transformation
    b. transduction
    c. conjugation
    d. all of the above require direct contact between donor and host

11. The genetic code is called universal because:

    a. all procaryotes use the same code
    b. all eucaryotes use the same code
    c. all mammals use the same code
    d. all living things use sthe same code

12. Rolling circle DNA synthesis occurs in___.

    a. eucaryotic chromosomal DNA synthesis
    b. eucaryotic chromosomal RNA synthesis
    c. procaryotic plasmid DNA synthesis
    d. procaryotic plasmid RNA synthesis

13. Uncorrected errors in DNA replication become___.

    a. replication forks
    b. mutations
    c. lagging strands
    d. gyrases

14. Which of the following statements is FALSE, concerning DNA?

    a. The two strands of DNA are in anti-parallel orientation to each other
    b. The backbone of DNA is made up of sugars and phosphates linked by ionic bonds
    c. The rungs of the DNA ladder are made up of complementary base pairs
    d. DNA replication is semi-conservative

15. If you see a sequence of single stranded nucleic acid with uracil and no thymine, you are looking at _____.

    a. plasmid DNA
    b. chromosomal DNA
    c. RNA, not DNA
    d. Okazaki fragments

Chapter 10 Genetic Engineering

Building Your Knowledge

1) What is genetic engineering?

2) Give 2 examples of how we can use genetic engineering in society.

    a. _____

    b. _____

3) How do scientists cut DNA?

What types of sequences do these "molecular scissors" recognize?

Why is the creation of "sticky ends" important when cutting DNA?

4) How is ligase used in genetic engineering?

5) What is cDNA and how is it made?

   Does cDNA have introns?

6) What is gel electrophoresis?

   Why do DNA molecules move toward the positive pole of a gel?

   Which move faster in a gel, large pieces of DNA or small pieces?

   Are the slow-moving pieces found at the top or the bottom of the gel?

7) What are gene probes and what can they be used for?

   Would gene probes work if the complementary base pairs of DNA did not exist? Why or why not?

8) Diagram the process of Sanger DNA sequencing. How many tubes do you set up? Why?

   Why do dideoxy-nucleotides stop the elongation of DNA?

9) If you were going to set up a test-tube for PCR, what would you add?

10) Why is it important that you use a thermostable enzyme for PCR?

How are PCR products analyzed?

11) What is recombinant DNA?

12) Diagram the process of cloning. Label the insert DNA, vector and cloning host.

How are clones containing the gene of interest separated from those that don't have the recombinant plasmid?

13) Cloning vectors must have 3 key traits. Name them.

a. _____

b. _____

c. _____

14) What are some of the desirable features of cloning hosts?

a. _____

b. _____

c. _____

d. _____

15) Why are recombinant proteins, such as insulin, better medicines than animal or human-derived proteins?

What proteins have been produced and are available for medicinal use?

16) What are transgenic organisms?

17) What is Frostban and how does it work?

18) How are genetically modified organisms regulated by the government?

19) What is bioremediation?

20) Give 3 examples of genetically engineered plants, listing the gene inserted and the advantage of having the new gene.

Plant	Gene Inserted	Advantage

21) Give 4 examples of genetically engineered animals, listing the gene alterations and the reason for these alterations.

Animal	Gene Alteration	Reason for Alteration

22) What is gene therapy and how does it differ from recombinant protein production?

23) What was the first human disease treated with gene therapy? When was the clinical trial? Was it successful?

24) How is germline therapy different from somatic cell gene therapy?

Which form of gene therapy corrects gene defects for an individual, and future generations?

25) What is the goal of antisense gene therapy?

Which step of protein synthesis does it stop? _____

Do these therapies generally have more or fewer side effects than traditional antivirals? Why or why not?

26) What is triplex DNA? What step of DNA synthesis does this process stop?

27) Compare and contrast linkage maps, physical maps and sequence maps of genomes. Which are the most detailed?

28) Does most of the DNA in the human genome code for proteins?

29)How is DNA technology used in forensic science? When was it first used and where?

30)How has PCR improved the use of DNA fingerprinting by forensic scientists?

31)How can DNA fingerprinting be used in human clinical genetics?

What genetic diseases can be detected by DNA fingerprinting?

Organizing Your Knowledge

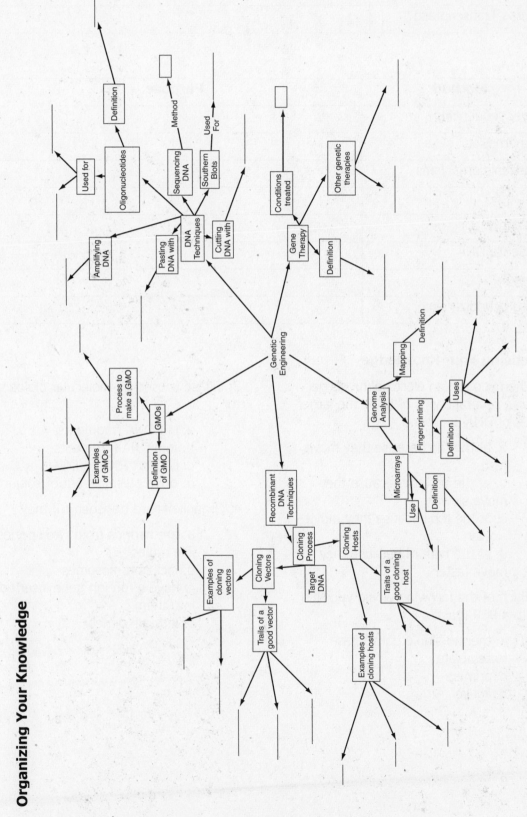

Enzyme	Use in Nature	Use in Lab
Restriction Endonucleases		
Ligase		
Reverse Transcriptase		

Method	Purpose
Sanger sequencing	
Southern Blot	
Transgenic organisms	
PCR	
Microarrays	
Ant-Sense DNA	
Cloning	
In situ hybridization	

Practicing Your Knowledge

1. When you run an electrophoresis gel and see several bands of DNA, the larger pieces of DNA are found _____.

    a. at the top because they move faster
    b. at the bottom because they move slower
    c. at the top because they move slower
    d. at the bottom because they move faster

2. Which of the following would you NOT add to a PCR reaction?

    a. thermo-stable polymerase
    b. template
    c. primers
    d. ligase

3. Ligase is used by molecular biologists to___.

    a. cut DNA fragments
    b. join DNA fragments
    c. convert RNA to DNA
    d. drive plasmids out of solution

4. Organisms are transgenic if they ___.

    a. are hybrids from two species - like mules
    b. lack chromosomes
    c. have a foreign gene inserted into them
    d. lack all genes

5. Restriction endonucleases recognize ___.

       a. mis-matched DNA segments and repair them
       b. palindromic DNA sequences and cut them
       c. mis-matched DNA sequences and cut them
       d. palindromic DNA sequences and repair them

6. A plasmid is -

       a. found only in fungus
       b. a bit of chromosomal DNA
       c. linear DNA found in prokaryotes
       d. circular DNA separate from chromosomal DNA

7. In the US, releases of recombinant microbes must be approved by ___.

       a. FDA
       b. EPA
       c. FBI
       d. NIH

8. Cloning of DNA fragments and inserting them into a host forms ___.

       a. vectors
       b. cloning hosts
       c. recombinant organisms
       d. PCR reactions

9. PCR is used to ___ .

       a. amplify DNA segments
       b. cut DNA segments
       c. join DNA segments
       d. force DNA segments into host cells

10. Anti-sense DNA will bind to ___ and prevent translation.

       a. tRNA
       b. mRNA
       c. ribosomes
       d. rRNA

11. Recombinant human proteins are often better than animal products or human products because ___.

       a. infectious diseases can be spread through these proteins
       b. humans often develop allergies to animal proteins
       c. there often isn't enough protein to treat all the patients that need a given protein.
       d. all of the above

12. The steps of the PCR cycle are ___ (in order).

       a. denaturation, priming, extension
       b. cutting, denaturation, priming
       c. priming, denaturation, cutting
       d. extension, denaturation, priming

13. Dideoxynucleotides are used for ___

       a. cutting DNA
       b. joining DNA fragments
       c. amplifying DNA
       d. sequencing DNA

14. cDNA is made by ___.

       a. incubating DNA with RNA polymerase
       b. combining plasmid and target DNA
       c. incubating RNA with reverse transcriptase
       d. cutting DNA probes

15. Good cloning hosts need all of the following traits EXCEPT:

       a. fast growth rate
       b. well-known genome
       c. pathogenic
       d. maintenance of foreign DNA through multiple generations

Chapter 11 Physical and Chemical Control of Microbes

Building Your Knowledge

1) What is the difference between physical and chemical methods of microbial control?

Method of Control	Definition	Examples
Physical		
Chemical		

2) How do sterilization and disinfection differ?

Can something be almost sterile? Explain.

3) An antimicrobial agent can be –cidal or –static (e.g. bactericidal or fungistatic) What is the difference between a –cidal agent and a -static agent?

If you take bacteriostatic antibiotics for an infection, what will happen when you stop taking the antibiotic?

4) Differentiate between the use and definition of antiseptics and disinfectants.

Agent	Used On	Examples
Disinfectant		
Antiseptic		

5) Your bathroom in a public restroom may say "sanitized for your protection". Is sanitization the same as sterilization?

6) When is a microbe considered dead?

7) How do the number of organisms and their state of growth (spores vs vegetative cells) affect microbial death rate?

8) How does agent concentration affect microbial death rate?

9) What are the four major cellular targets for antimicrobial agents?

        a. _____

        b. _____

        c. _____

        d. _____

10) How does heat affect microbial cells?

        a. Moist heat -

        b. Dry heat –

11) How do surfactants work?

12) Why does radiation limit microbe growth?

        a. Ionizing radiation-

        b. Non-ionizing radiation -

13) Name one agent that targets bacterial cell walls.

14) Why is cold not used to kill microbes?

15) Which is more effective in killing microbes, moist heat or dry heat? Why?

   Which does an autoclave employ?

16) Does pasteurization sterilize milk?

   What does it do?

17) How is the usefulness of non-ionizing radiation limited?

18) Are more toxic compounds generally better at killing microbes or worse?

19) How does bleach work as a disinfectant?

20) How does alcohol kill bacteria?

21) What is the cellular target of detergents?

22) What are the cellular targets of heavy metals?

23) Formaldehyde is used to sterilize contaminated areas – how does it kill microbes?

26) If you had a non-autoclave safe object that you needed to use as a surgical implant, how would you clean it for use? Discuss several options and the advantages of each.

Option A _____

Advantages -

Disadvantages -

Option B _____

Advantages -

Disadvantages -

Option C_____

Advantages -

Disadvantages -

Organizing Your Knowledge

Control Method	Type (Physical, Chemical, Mechanical)	Cellular Target	Example
Moist Heat			
Dry Heat			
Ethylene Oxide			
Pasteurization			
Halogenation			
Incineration			
Ionizing Radiation			
Non-ionizing radiation			
Surfactants			
Hydrogen Peroxide			
Quaternary Ammonia Compounds			
Ethylene Oxide			
Heavy Metals			
Aldehydes			
Sanitizing			
Cold Temperatures			
Phenolics			

Practicing Your Knowledge

1. Cooking utensils are often _____ between uses in restaurants.

    a. sanitized
    b. irradiated
    c. sterilized
    d. autoclaved

2. Dry heat is more effective at killing microbes than moist heat

    a. True
    b. False

3. Which of the following is NOT a physical agent used to control microbes?

    a. moist heat
    b. dry heat
    c. radiation
    d. ethylene oxide

4. 100% Ethanol solutions are more effective antimicrobial agents than are 70% solutions.

    a. True
    b. False

5. Alcohol sterilizes skin.

    a. True
    b. False

6. Non-ionizing radiation, such as _____ kills bacteria by _____

    a. UV: denaturing lipids
    b. X rays: denaturing proteins
    c. UV: inducing mutations in DNA
    d. X rays: inducing mutations in RNA

7. Household bleach is used as a common disinfectant because it has _____.

    a. hypochlorite
    b. idonic
    c. phenol
    d. chlorohexidine

8. Placing objects in boiling water _____ them.

    a. sterilizes
    b. disinfects
    c. sanitizes
    d. tyndallizes

9. Which of the following has the HIGHEST resistance to killing?

    a. naked viruses
    b. vegetative bacterial cells
    c. bacterial endospores
    d. viral endospores

10. A chemiclave sterilizes objects by exposing them to _____ gas.

    a. formaldehyde
    b. surfactant
    c. ethylene oxide
    d. silver tincture

11. Which of the following is NOT a chemical agent used to control microbes?

    a. filtration
    b. phenol
    c. halogens
    d. glutaraldehyde

12. An autoclave sterilizes objects by using _____.

    a. chemical treatments
    b. multiple low-heat treatments
    c. steam under pressure
    d. multiple incinerators

13. Chemical surfactants kill microbes by _____.

    a. damaging their DNA
    b. stopping ribosomal movement
    c. denaturing proteins
    d. disrupting membranes

14. A viricidal agent will ___.

    a. kill bacteria
    b. kill viruses
    c. stop bacterial growth, but not kill the bacteria
    d. stop viral growth, but not kill the viruses

15. Treatment with cold _____

    a. slows microbial growth
    b. kills all microbes
    c. can be used to disinfect surfaces
    d. kills all fungus

Fun with Your Knowledge

Control of Microbes Terminology

```
N G O G S H M S H E G I P B I C Y Z J Q I D F P V I I Q O N
E H E G U C R T I N C T U R E S Y I I E E M Z M C Q V F F X
P V R W R Y W R A V O A U T O C L A V E Z X Q T W C F Y U E
F J Z B X W Y M E U Z P H I F V D V V T G W A Z M R B A B D
A S E P S I S X O H N F B H M M Q M U S H F I P S R A R M E
C X G G J N O V S K T H M Y I P Y H W P M F P Q X S C L T Q
P A S T E U R I Z A T I O N O L I G O D Y N A M I C T N H U
U L L X B X N X X A X Q H C L A J U A A C M B Q B J E E C S
X N U L E D Z W K D J D N Z A E N B N N S D A P Z Q R G S V
D Q G X L J S R Y T J I A Z O R K T N H M W O O T S I B J J
L Z I J W Y B P A N Y P L T Y A S Q I E U K V G B O C I U F
W P W A P V Q I D F X B B J M C N B Y S B Q O Y Z K I F T I
U X M P L S M Z D Q M O Q M R J V Z M V E L G P A D D N G C
Y I B E R Y A B E T H Y L E N E O X I D E P C D U Y E Q O S
X I R G E Z Y A R T K K E B D L W B X Q B Q T B V D M A D T
J G S S L N U C R L B J S U R F A C T A N T S I W V G L G E
N L B U Y J R T O W L B C O C S P O R I C I D E C U U C I R
K T X Q A H C E D E N A T U R E A V P P A N C X L S T S Q I
S S I W O D Q R F I L T R A T I O N M S U M P A S E K K A L
K Y X U K A F I P C K B R J R Q F X D E M F T J I Y S Q K E
O E O G C M E O D N A H H D O T Y E V Y B E O N R X L J H D
D V B D O G J S M V M U M V T D R G J W H K L Z S K K B T L
E P T H X X M T I O N I Z I N G C I E T H A N O L L M N M Q
D X J Q R Y M A K K J W Q H K P C N C Y J Y M F A N S T B S
I A K A V M K T P E F N B X I G G N R L S K J Y Y Z N T N F
T G O T X X U I C D A E O T E W Q T Z E O Z M I A C S L O E
Q F B R V U S C C F G I T H S J B J P V S S O O T S Y H K W
M V V D J W P B Y K G K I F I Y I T W G N X A D E A T H S K
I M W O T N Y K I I J G T R D X L M P R H B I N P C Z S E F
Z Q I F P A S R Z F H R D B N G C M H W V W B F D J O E E A
```

1. (___) cellular target of ionizing and nonionizing radiation

2. (_____) chemical agents used directly on body surfaces to destroy or inhibit pathogens

3. (_____) practices that prevent entry of infectious agents into sterile tissues and prevent infection

4. (_____) device using steam under pressure to sterilize objects

5. (_____) chemical that destroys bacteria except for endospore

6. (_____) agents that prevent the growth of bacteria are called ___.

7. (_____) moist heat causes proteins to ____ or denature

8. (_____) microbes are ____ if they have permanently lost the ability to grow, even under ideal conditions

9. (_____) disruption of proteins, making them non-functional

10. (_____) chemical agent found in mouthwash and waterless hand sanitizers

11. (_____) chemical sterilant that is used in "chemiclaves"

12. (_____) sterilization technique that strains fluid through openings too small for microbes to pass through

13. (_____) radiation that ejects electrons from their orbital shells, creating ions.

14. (_____) heavy metals are effective antimicrobial agents at very low concentrations and are therefore called

15. (_____) heating liquids to kill potential infection and spoilage agents

16. (_____) agent capable of destroying bacterial endospores

17. (_____) material that is devoid of all viable organisms including viruses

18. (_____) microbiocidal detergents that lower the surface tension of cell membranes

19. (_____) chemical agents dissolved in alcohol are called ___.

20. (_____) chemical agent found in most antibacterial soaps

Chapter 12 Drugs, Microbes, Host

Elements of Chemotherapy

Building Your Knowledge

1) In 1900, what percentage of all children in the United States died of infectious disease before age 5?

2) Explain the concept of selective toxicity.

3) Design the perfect antibiotic. Please consider ease of administration, target microbe, distribution throughout the body (if necessary for the targeted microbe) length of activity, host toxicity and potential for the development of antibiotic resistance to the drug.

   Does the perfect antibiotic actually exist? Why or why not?

4) Compare and contrast narrow and broad spectrum antibiotics.

   If you know what the causative agent of a disease is, which would you want to use?

   If a patient is septic (seriously ill) with an unknown organism, which would you recommend?

5) Differentiate between chemotherapeutic and prophylactic drug use.

   Is taking penicillin for a case of Strep throat prophylactic or chemotherapeutic?

6) Where do most antibiotics originate, in the lab or from nature?

Why is this the case?

7) There are three interacting factors involved in the outcome of antimicrobial therapy. Name them.

   a. _____

   b. _____

   c. _____

8) Which drugs are most selectively toxic to bacterial cells (in general)?

Which are the least selectively toxic?

Why?

9) Are drugs that target the cell wall more or less selectively toxic than those that target the plasma membrane? Why?

10) Are penicillin and penicillin-like antibiotics more effective against actively growing cells, or old, dormant cells? Why?

Are they more effective against Gram positive or Gram negative cells? Why?

11) What is competitive inhibition and what does it have to do with sulfonamide activity against bacterial cells?

12) Why don't sulfonamide's damage host cells as much as they do bacterial cells?

13) How are chloroquine and AZT similar?

How are they different?

14) If both eucaryotes and procaryotes have ribosomes, why are antimicrobial that target ribosomes selectively toxic?

15) Are the membrane-disrupting drugs generally used topically (on body surfaces) or administered internally?

16) Fill in the diagram with several example antibiotics that target each of the following structures or processes in a bacterial cell.

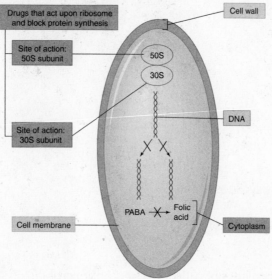

17) How is most penicillin produced?

18) List 3 members of the penicillin family.

a. _____

b. _____

c. _____

19) What is the advantage of using semi-synthetic penicillins, like ampicillin or nafcillin?

20) Why add clavulanic acid to penicillins (e.g. Augmentin)?

21) What are the major problems that limit the usefulness of the penicillin antimicrobials?

22) What are cephalosporins?

How are they generally administered?   Why?

What are the "generations" of cephalosporins based on?

What are the major drawbacks of cephalosporin use?

23) What are the tetracyclines?

What is their mechanism of action?

24) Why is chloramphenicol not a widely used antimicrobial?

25) What do erythromycin, clindamycin, vancomycin and rifamycin have in common?

How do these drugs act against bacteria?

Why are they not more widely used?

26) What are the bacillus antibiotics?

What is their mechanism of action?

Why are they generally not given systemically?

27)Where do sulfamides come from?

How does this origin differ from that of penicillins and cephalosporins?

28)What are fosofmycin and synercid?

Why are they not widely used?

29)Why are scientists hopeful that resistance to the oxazolidinones will be slow to develop?

30)Why are anti-fungals generally more toxic to human tissues than antibacterial agents?

31)List 4 separate antifungals and the conditions they treat.

Why are polyenes effective against fungal cells, but not bacterial cells?

32)Why are there few effective anti-parasite drugs?

What drugs are used to treat malaria?

What drugs are used to treat roundworm?

33)Why is selective toxicity so difficult to achieve in anti-viral therapies?

34)Why are viral diseases like measles and mumps fairly rare in the US?

35) What are 3 basic mechanisms of action of anti-viral agents?

      a. _____

      b. _____

      c. _____

36) Why is the fact HIV is a retrovirus significant when designing anti-viral therapies?

37) What drugs are commonly used to treat HIV and how do they inhibit the viral cycle?

38) How does chromosomal drug resistance originate? Does this type of resistance spread in a population?

39) What is the difference between intrinsic and extrinsic drug resistance? Which is of more concern?

40) What are R factors? Do these spread through a bacterial population?

41) Describe 4 distinct ways bacteria may become resistant to antibiotics they were once sensitive to.

      a. _____

      b. _____

      c. _____

      d. _____

42) What are beta-lactamases and what antimicrobial drugs do they confer resistance to?

43) Do pumps generally confer resistance to 1 type of antimicrobial or many? Why?

How do bacteria become resistant to rifampin or streptomycin?.

44) How do bacteria become resistant to sulfonamide?

45) Does exposure to an antibiotic increase or decrease the percentage of resistant cells in a population? Explain.

46) Why does combining drug therapies limit the spread of drug resistance?

47) There are 3 major categories of antibiotic side-effects. Name them.
      a. _____

      b. _____

      c. _____
48) Why are the liver and kidneys often damaged by antibiotics?

49) Why are tetracyclines not given to pregnant women or young children?

50) How may antibiotics cause diarrhea? (list 2 ways)

51) If a person takes penicillin once and does not have an allergic reaction to it, does that mean they are not allergic? Explain.

52) What 3 factors do doctors generally consider when choosing antimicrobial therapies?

    a. _____

    b. _____

    c. _____

53) If the causative agent of a disease is unknown, do doctors generally give narrow-spectrum or broad-spectrum antibiotics? Why?

54) What 2 methods are commonly used to tell which antibiotics are most and least effective against a given pathogen?

    a. _____

    b. _____

55) Diagram a Kirby-Bauer plate. Draw the plate, antibiotic disks and zones of inhibition.

If Drug A has a larger zone than Drug B and both drugs are at the same concentration and are the same size, which drug A or B is more effective?

56) What is the MIC and how is it used?

57) What is the therapeutic index (TI)?

If a drug has a TI of 5 is it more or less safe to use than one with a TI of 0.5? Why?

58) What other variables do physicians need to be concerned with when choosing antimicrobial drugs?

59) What current practices are contributing to the mis-use and spread of drug resistance in microbial populations?

Organizing Your Knowledge

Antimicrobial agent	Mechanism of Action	Commonly used to treat
penicillin		
sulfonamide		
chloroquine		
gentamicin		
polymyxins		
nystatin		
rifampicin		
mebendazole		
metronidazole		
ribavirin		
AZT		
amphotericin B		
chloramphenicol		
tetracycline		
vancomycin		

Antimicrobial Drug	Group	Mechanism of Action	Mechanism of Resistance
vancomycin			
tetracycline			
sulfonamide			
rifampicin			
ribavirin			
polymyxins			
penicillin			
nystatin			
metronidazole			
mebendazole			
gentamicin			
chloroquine			
chloramphenicol			
AZT			
amphotericin B			

Practicing Your Knowledge.

1. Which of the following is NOT a common target for antibacterial drugs?
    a. cell wall synthesis
    b. nucleic acid structure
    c. protein synthesis
    d. bacterial cell nucleus

2. All of the following antibiotics target procaryotic ribosomes EXCEPT___.
    a. streptomycin
    b. cephalexin
    c. gentamicin
    d. erythromycin

3. Prophylatic antibiotics are given ___
    a. after a person is infected with a virus
    b. to people at increased risk of viral infection
    c. after a person is infected with bacteria
    d. to people at increased risk of bacterial infection

4. An antibiotic with a high therapeutic index (TI) ___.
    a. is a less risky choice than one with a low TI
    b. is generally very toxic
    c. has a high MIC and low toxic dose

5. Which of the following is NOT a characteristic of an ideal antimicrobial drug?
   a. not excessive in cost
   b. microbistatic, not microbicidal
   c. selectively toxic to microbe
   d. easy to deliver to site of infection

6. Anti-viral drugs are ____.

   a. commonly used to treat head colds.
   b. hard to design because the viruses are intracellular parasites
   c. generally safer to use than anti-bacterial drugs
   d. not subject to anti-viral resistance

7. A MDR pump will confer resistance to ____.

   a. a single class of antibiotics (e.g. the penicillins)
   b. many different antibiotics from different groups
   c. only gram-negative bacteria
   d. only gram-positive bacteria

8. ____ is an example of a synthetic antimicrobic drug.

   a. polymyxin
   b. rifamycin
   c. tetracyline
   d. sulfonamide

9. When the cause of a disease is unknown, but suspected to be bacterial, a useful course of action would be ____.

   a. to start anti-viral therapy
   b. to disinfect the patient
   c. to start a broad-spectrum antibiotic
   d. to start a narrow-spectrum antibiotic

10. Two ways to determine an organism's resistance to antimicrobial drugs are ____ and ____ methods.

   a. MIC and therapeutic index
   b. Kirby-Bauer and therapeutic index
   c. MIC and Kirby-Bauer
   d. Kirby-Bauer and beta-lactamase

11. An organism becomes resistant to penicillin when it____.

   a. produces thymidine kinase
   b. acquires its folic acid from the enviroment
   c. produces beta lactamase
   d. loses its DNA

12. Clavulanic acid is added to the penicillin group of drugs because ____.

   a. it inhibits beta-lactamase enzymes
   b. it works against Gram-positive bacteria, penicillins don't
   c. it lengthens the shelf-life of penicillins
   d. it disrupts bacterial cell membranes

13. The major drawbacks to penicillin use are ____.

   a. development of resistance and host cell toxicity
   b. synergistic effects with anti-viral therapies
   c. development of resistance and host allergic responses
   d. purine degradation and host allergic responses

14. Antibiotics that disrupt microbial plasma membranes ____.

   a. are more toxic than those that disrupt microbial cell walls
   b. are commonly given systemically
   c. are not toxic to host cells
   d. generally have a high therapeutic index

15. ____ is commonly used to treat fungal infections.

   a. Tetracycline
   b. Vancomycin
   c. Amphotericin B
   d. Quinine

Chapter 13 Microbe-Host Interactions: Infection and Disease

Building Your Knowledge

1) What are the three basic outcomes of an encounter between microbes and humans?

   a. _____

   b. _____

   c. _____

2) What are the possible outcomes following microbial infection?

   a. _____

   b. _____

   c. _____

3) Once a person develops a disease, what can happen to the individual?

   To the infecting microbe ?

4) Do all infections lead to disease? Why or why not?

5) Give 2 specific examples of how normal flora may protect against infection.

   a. _____

   b. _____

6) In general, which sites of the human body lack normal flora?

7) Why do many doctors recommend eating yogurt if you have the stomach flu?

8) Name two sites that have significant numbers of normal flora.

9) Describe how newborn infants acquire normal flora.

10) How are transient and resident flora similar to one another?

How are they different?

11) What is an endogenous infection? Where do the infecting bacteria come from?

Which people are at particular risk for endogenous infections?

12) Certain strains of *Streptococcus* and *Neisseria* are normal flora, while others are pathogenic. What trait do the pathogenic strains share and non-pathogenic strains lack?

13) Are rough strains or smooth strains of bacteria generally more pathogenic?

Which ones have a capsule?

What does a capsule do for a microbe?

14) Why do scientists study normal flora, if it doesn't cause disease?

What are gnotobiotic animals?

How do they differ from "normal" animals?

15) How do pathogens differ from normal flora?

16) Which are more dangerous, pathogens at biosafety level 1 or biosafety level 4?

17) What are virulence factors?

18) Describe 3 separate virulence factors and how they contribute to the disease process.

      a. _____

      b. _____

      c. _____

19) How is the portal of entry important to the spread of disease? If a person has influenza virus on his or her hands, will that person get influenza? Explain.

20) Describe 3 separate portals of entry and the pathogens that enter the body through them.

      a. _____

      b. _____

      c. _____

21) Define infectious dose. Would you rather drink 1000 *V. cholera* cells or inhale 10 *M. tuberculosis* cells? Why?

22) Why is adhesion of particular importance to pathogens?

How may viruses adhere to cells?

How may spirochetes?

Other bacteria?

23) How are an exoenzyme and exotoxin similar?

How are they different?

Give 3 specific examples of exoenzymes that are virulence traits.

24) Differentiate between an intoxication and infection. Which one requires active growth of bacteria?

25) Differentiate between exotoxins and endotoxins.

Give 3 specific examples of exotoxins as virulence traits and how they contribute to disease processes.

26) Give three specific examples of how pathogens may evade an immune response.

a. _____

b. _____

c. _____

27) A focal infection has characteristics of both a localized and systemic infections. How is this the case? (define localized, systemic and focal infections).

28)How are a mixed infection and a primary-secondary infection similar?

How are they different?

29)Is the common cold an acute or chronic disease? Why?

30)Differentiate between the signs and symptoms of an illness.

Is a fever a sign or a symptom?

Is a headache a sign or a symptom?

What is a syndrome?

31)How are viremia, bacteremia and septicemia similar?

How are they different?

32)Describe 4 separate portals of exit and the pathogens that use them.

a. _____

b. _____

c. _____

d. _____

33)What are latent infections? Give 2 examples of latent infections.

34)What are the sequelae of Strep throat, Lyme disease and polio?

35)Which branch of microbiology studies the effects of disease in a population?

36)What are reportable diseases?

37)Distinguish between the prevalence and incidence of a disease?

If there are 20 cases in a population of 100, what is the prevalence of the disease?

If the following week there are 10 more new cases, what is the incidence?

38)Differentiate between morbidity and mortality.

Morbidity –

Mortality –

If 40 individuals in a population of 500 are sick with viral infection, what is the morbidity rate? _____

If 5 of those 40 then die, what is the mortality rate? _____

39)If a disease is endemic in an area, what does that indicate?

40)Differentiate between an epidemic and pandemic.

41)What is the "iceberg effect" when applied to epidemiology?\

42) Differentiate between a reservoir and a source of infection.

43) What is the difference between asymptomatic carriers and passive carriers?

44) Houseflies and roaches can transmit disease because they are _____ vectors.

45) The transmission of malaria by mosquito bite is an example of a _____ vector.

46) Are humans the natural host of rabies and West Nile viruses Explain.

47) How may an infectious disease be spread from person to person?

How do direct and indirect contact methods of transmission differ?

Give examples of each.

48) What is a fomite?

49) Is tetanus a communicable infectious disease? Explain.

50) Differentiate between aerosol spread and droplet nuclei spread of respiratory pathogens.

Which spread the hardier pathogens?

51) Where are nosocomial infections acquired?

52) How do you determine if a particular microbe is the causative agent of a given disease? (list and describe all four steps of Koch's Postulates).

       a. _____

       b. _____

       c. _____

       d. _____

Has this process been done for ALL known infectious diseases? Why or why not?

Organizing Your Knowledge

Entry Point	Pathogen	Disease	Exit
		cholera	
	S. pyogenes		
		pinkeye	
	polio virus		
		giardiasis	
	M. tuberculosis		
		AIDS	

Area of the Body	Normal Flora (yes/no)	Internal or External?
Skin		
Heart		
Upper Respiratory Tract		
Bones		
Mouth		
Muscles		
Vagina		
Ovaries		
Urine (in bladder)		
Lungs		
Gastrointestinal Tract		
Saliva (in salivary glands)		
External Eye		
Blood		

Practicing Your Knowledge

1. A pathogen may be ___.

    a. a fungus
    b. a bacteria
    c. a virus
    d. all of the above

2. Which of the following organisms enters the body via the gastrointestinal tract?

    a. Rabies
    b. Mycoplasmas
    c. Polio virus
    d. Histoplasmas

3. In general, sites of the human body devoid of normal flora are _____.

    a. non-existant
    b. external body surfaces
    c. internal body surfaces, such as the intestinal tract
    d. internal organs, such as the bladder

4. An indirect method of disease transmission would be by ___.

    a. rabid dog bite
    b. mosquito transmission of malaria
    c. touching a fomite
    d. kissing someone with mononucleosis

5. Two of the heaviest areas of microbial growth are

    a. the intestine and bladder
    b. the intestine and mouth
    c. the skin and bladder
    d. the mouth and bladder

6. Nosocomial infections are acquired _____.

    a. by direct contact with an animal
    b. in hospitals
    c. though sexual activity
    d. by indirect contact with animal feces

7. Which of the following is a sign of an active infectious disease?

    a. fever
    b. fatigue
    c. headache
    d. sore throat

8. Which of the following is NOT an adhesion factor for pathogens?

    a. capsules
    b. flagella
    c. fimbriae
    d. cell walls

9. The transient population of flora _____.

    a. normally grows on humans
    b. is found on deeper layers of human skin and forms a stable population
    c. is acquired by routine contact
    d. is a stable population that causes disease

10. The _____ of a disease is the ratio of the number of new cases to the number of healthy people in a given population.

    a. incidence
    b. prevalence
    c. morbidity
    d. mortality

11. If a pneumonia-causing microbe causes pneumonia, then worsens to cause septicemia we would call this a _____ infection.

    a. toxemic
    b. localized
    c. focal
    d. mixed

12. The portal for the greatest number of pathogens is the ___.

    a. respiratory tree
    b. gastrointestinal tract
    c. skin
    d. urogenital (STDs)

13. Endogenous infections are caused by ___.

    a. true pathogens causing disease
    b. fungal infections only
    c. normal flora
    d. viral infections only

14. Gnotobiotic animals

    a. lack normal flora or have well-defined flora
    b. have a well-developed immune system
    c. often have more cavities than normal animals
    d. are less sensitive to gut pathogens like Salmonella

15. Intoxications are due to _____.

    a. the ingestion of live organisms that grow and produce endotoxin
    b. the ingestion of exotoxins
    c. the growth of toxin-producing bacteria in the blood
    d. the ingestion of live organisms that grow and cause disease

Chapter 14 The Nature of Host Defenses

Building Your Knowledge

1) List the three "lines of defense" a host organism uses to prevent invasion by pathogens.

   a. _____

   b. _____

   c. _____

   Which of these defense mechanisms are innate defenses?

   Which of these are acquired through exposure to a pathogen or vaccine?

2) List the physical barriers to infection in a mammalian body.

3) Humans continuously produce and lose skin cells. How does this aid in the prevention of infection?

   What other barriers to infection does the skin pose?

4) Most symptoms of the common cold are attempts from the body to rid itself of a virus. List the symptoms of a cold and their effects on viral infection.

Symptom	Effect on Virus

5) Chemical defenses are chemicals produced by the body to impede the growth of microbes. List three of these defenses, how they impede the growth of microbes, and where they are produced.

Chemical Defense	Anti-microbial action	Location Produced

6) Genetic defenses may exist between different species (interspecific) or within a single species (intraspecific). Give an example of each.

    a. Interspecific genetic defense -

    b. Intraspecific genetic defense -

7) What are the 3 major tasks the immune system accomplishes for a healthy body?

    a. _____

    b. _____

    c. _____

8) Describe the structure and function of the reticuloendothelial system. Where is the RES in a human body?

How is the RES connected to the blood system, extracellular fluid, and lymphatic systems?

9) What is hemopoiesis?

Where does it take place in infants?

Where does it take place in adults?

10) Granulocytes and agranulocytes are two groups of cells produced. How are they similar?

11) List the three types of granulocytes and the two types of agranulocytes and the function of each cell type?

Cell	Cell type	Function

   Which are the most numerous?

12) How are monocytes and macrophages related?

13) What is the structure and function of dendritic cells?

14) How are mast cells and basophils related?

15) How are platelets produced and what is their function in the body?

16) Chemotaxis and diapedesis are related processes. How are they similar? How are they different? How are they related?

17) What is the lymphatic system, where is it found and what purpose does it serve? List and describe the lymphoid tissues.

Lymphoid Organ	Location	Function
Vessels		
GALT		
Lymph Nodes		
Spleen		
Thymus		
Tonsils		

18) What are the 3 basic functions of an inflammatory response?

a. _____

b. _____

c. _____

19) What are the four cardinal (classic) signs of inflammation, what are the Latin terms for these signs and what causes these symptoms?

Cardinal Sign	Latin Term	Cause of the Symptom

20) What cells are generally the first to reach the site of inflammation?

How do they get there?

Are these cells part of the specific or non-specific response to invasion?

21) What are the next cell types into an area of inflammation?

Are they specific or non-specific?

What is their function?

22) What is pus and what do we call bacteria that cause pus to form?

23) What are the possible causes of a fever?

How does a fever help fight an infection?

24) Which white blood cells (leukocytes) are phagocytes?

Which are not?

How do phagocytes kill bacteria?

25) What is interferon and how does it limit the spread of viruses?

What other things do interferons contribute to the immune response?

26) Differentiate between a substance that is pyrogenic and one that is pyogenic? Can something be both pyrogenic and pyogenic?

27) What is the complement cascade?

Describe the three major steps to the cascade.

There are two infection-fighting results of this cascade. What are they?

Organizing Your Knowledge

Immune Cell	Function	Phagocyte?	Granulocyte?
Neutrophils			
B cells			
Monocytes			
Eosinophils			
Mast cells			
Macrophages			
T cells			
Dendritic Cells			

Inflammation Event	Description	Cause of Event
Rubor		
Tumor		
Dolor		
Calor		

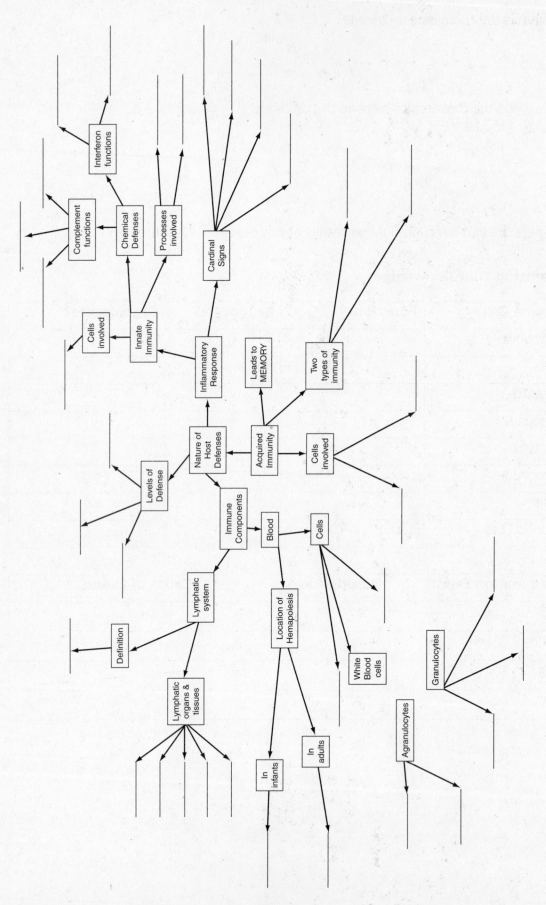

Complement functions

Interferon functions

Chemical Defenses

Processes involved

Cardinal Signs

Cells involved

Innate Immunity

Inflammatory Response

Leads to MEMORY

Two types of immunity

Nature of Host Defenses

Acquired Immunity

Cells involved

Levels of Defense

Immune Components

Blood

Cells

Lymphatic system

Location of Hemapoiesis

White Blood cells

Definition

Lymphatic organs & tissues

In infants

In adults

Agranulocytes

Granulocytes

Practicing Your Knowledge

1. Granulocytes include _____.

    a. macrophages and neutrophils
    b. monocytes and eosinophils
    c. basophils and lymphocytes
    d. neutrophils and basophils

2. B cells and T cells are ___.

    a. both lymphocytes
    b. both part of cell mediated immunity
    c. both capable of producing antibody
    d. both granulocytes

3. The non-specific defenses, such as phagocytes, are part of a body's ___ line of defense against infection

    a. first
    b. second
    c. third
    d. fourth

4. Cytokines aid in an immune response by ___.

    a. directly killing viruses
    b. activating leukocytes
    c. activating erythrocytes
    d. activating the complement cascade

5. Which of the body's fluid-filled spaces does NOT participate heavily in and immune response?

    a. bloodstream
    b. reticuloendothelial system
    c. lymphatic system
    d. cerebrospinal fluid

6. The lymphatic system includes all of the following EXCEPT ____.

    a. the spleen
    b. lymph nodes
    c. the heart
    d. the thymus

7. Any substance that causes fevers to develop is called a ____.

    a. cytokine
    b. pyogen
    c. pyrogen
    d. interferon

8. Which of the following cells is NOT part of the monocyte line of differentiation?

    a. monocytes
    b. macrophages
    c. platelets
    d. dendritic cells

9. Which of the following is an example of a chemical barrier to infection?

    a. blinking and lacrimation
    b. lysozyme
    c. lack of receptors on humans for distemper
    d. desquamation

10. Leukocytes that are phagocytic are ___.

    a. B cells and neutrophils
    b. platelets and B cells
    c. neutrophils and macrophages
    d. dendritic cells and B cells

11. Which of the following is mis-matched?

    a. rubor- redness
    b. tumor - swelling
    c. calor - heat
    d. dolor - pus formation

12. Which of the following is mis-matched?

    a. basophils-histamine
    b. neutrophils - specific immunity
    c. macrophages - phagocytes
    d. B cells - humoral immunity

13. A person with a high eosinophil count will likely ____.

    a. have an active helminth infection
    b. have an active histamine response
    c. have an active bacterial infection
    d. develop a pus-filled abscess

14. Human blood consists of ____.

    a. plasma, white blood cells, blue blood cells
    b. fibrin, plasma, white blood cells
    c. white blood cells, red blood cells, plasma
    d. plasma, lysozyme, hematin

15. Immune cells cross blood vessels to enter tissue spaces by _____.

    a. differentiation
    b. phototaxis
    c. complement
    d. diapedesis

Chapter 15 Specific Immunity

Building Your Knowledge

1) B cells and T cells are both _____.

2) Where do B cells mature?

Where do T cells mature?

3) B cells produce _____ in response to antigen.

4) Receptors are found on cells. What are 4 major functions of receptors?

5) A specialized group of receptors are MHC antigens. What does MHC stand for?

What are the 3 classes of MHC genes?

Which cells express MHC I and which cells express MHC II?

6) What is the clonal selection theory?

Do lymphocytes change their antibodies or other receptors to match a given antigen?

Do lymphocytes that recognize a given antigen already exist, but multiply upon exposure to their antigen?

7) What is tolerance and how does it occur?

8) What are immunoglubulins and which cells produce them?

9) Label the following figure, indicating the location of 3 disulfide bonds, heavy chains, light chains, variable regions, constant regions, the antigen binding site, the Fab area and the Fc region.

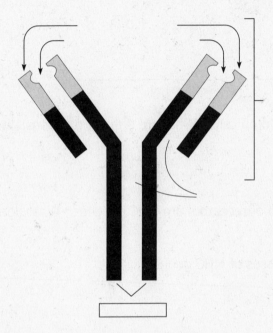

10) How is diversity generated in antibody molecules during B cell development?

What genes are involved in making a heavy chain of an antibody molecule?(list 4)

How many options are there for each gene needed?

What genes are needed to make a light chain?

11) How are T cell receptors similar to immunoglobulins?

How are they different?

12) Where do B cells mature in birds?

Where do they mature in humans?

13) Where do B cells go after maturation?

Do B cells circulate or "home" to a particular region?

14) Where do T cells mature?

Do T cells circulate in the blood or home to a specific area?

15) What is an antigen?

In general, what traits make a good antigen rather than a poor antigen?

What is a hapten?

How do you produce antibodies against a hapten?

16) Differentiate between auto-antigens and allo-antigens.

Which are of concern to transplant specialists?

17) What are superantigens and which group of lymphocytes do they directly stimulate?

18) How do most antigens enter the body?

Where are they gathered up and concentrated after entry?

19) What are antigen presenting cells and how do they present antigen to the body?

Which T cells do antigen presenting cells (APCs) present antigen to?

Which MHC class is used by APCs to present antigen to these T cells?

20) If an APC presents a T cell-dependent antigen to a T-helper cell, how does the APC activate the T cell? How does the T cell activate a B cell?

21) What are the general characteristics of T-independent antigens?

22) Do B cells present antigen to T cells on their antibodies? Why or why not?

23) What signals do B cells require to activate?

Where do these signals come from?

24) What 3 things happen in a B cell response after activation?

1)

2)

3)

25) What may antibodies do to eliminate a pathogen? (list 4 specific things

a. _____

b. _____

c. _____

d. _____

26) List and differentiate between the 5 different classes of antibody.

Antibody Class	Description

Which is most prevalent in the blood?

Which is found on mucous membranes?

Which serves as a B cell receptor?

Which causes allergy?

Which is produced first, upon exposure to antigen?

27) Draw an antibody titer graph, with the primary and secondary response. Label the IgM curve, the IgG curve and the latent period.

Antibody Concentration

Time

28) A secondary immune response is stronger, longer and quicker than a primary response. Why?

29) Differentiate between monoclonal and polyclonal antibodies.

30) How are monoclonal antibodies made?

How are monoclonal antibodies used? (list 4 ways)

a. _____

b. _____

c. _____

d. _____

31) There are many different types of T cells list 4 and their function.

T cell subset	Function

32) Differentiate between active and passive immunization.

Which is quicker?

Which is longer lasting?

33) Differentiate between natural and artificial immunity.

Organizing Your Knowledge

MHC Class	Found on ____ cells	Presents antigen to:	Gets antigen from:
MHC I			
MHC II			

Molecules	Good antigens?	Poor antigens?
Exotoxins		
Repetitive structures		
Small Molecules		
Glycogen		
Glycoproteins		
Pure DNA		
Bacterial capsules		
Large Proteins		
DNA with protein		
Haptens alone		
Haptens with carriers		

Activation Event	B cells	T cells
Differentiate (to what?)		
Produce proteins (what proteins?)		
Clonal expansion (yes/no)		

Antibody Function	Fab-mediated?	Fc-mediated?
Agglutination		
Opsonization		
Neutralization		
Complement Fixation		
Allergen Response		
Binding to Mast Cells		

Trait	T_H	T_C	T_{DH}	T_S
Recognizes MHCII				
Has CD4				
Has CD8				
Recognizes MHC I				
Regulates Immune reactions				
Produces antibody				
Provides B cell and T cell help				
Produces perforins				
Causes late allergies				
Destroys virally infected cells				
Reduced in HIV/AIDS patients				

Attribute	T Cell Receptor	Antibody	Both
Antigen Binding Sites			
Variable and Constant Regions			
Light or Heavy Chains			
Formed by Genetic Modifications			
Secreted			
Binds MHC and Antigen			
Found on B cells			
Recognizes Free Antigen			
Recognizes Antigen bound to MHC			
Disulfide Bonds			

Practicing Your Knowledge

1. Which regions of an immunoglobulin bind to antigen?

a. variable regions of heavy chains, constant regions of light chains

b. variable regions of light chains, constant regions of heavy chains

c. variable regions of both chains

d. constant regions of both chains

2. Which of the following make good antigens?

a. starch, with its repetitive sugars

b. tetanus toxin, since its an exotoxin

c. hemoglobin, with its repetitive amino acids

d. oils, with the gycerol and fatty acids

3. B cells and T cells can make up the ___ immune response and are types of ____.

a. specific: lymphocytes

b. non-specific:neutrophils

c. specific:macrophages

d. non-specific:eosinophils

4. Immunotoxins are ___.

a. exotoxins produced by bacteria to kill immune cells

b. polyclonal antibodies that neutralize toxins

c. monoclonal antibodies that neutralize other antibodies

d. hybridized monoclonal antibodies that kill cancer cells

5. According to the clonal selection theory:

a. lymphocytes change to match antigen

b. naive lymphocytes are a diverse population

c. a lymphocyte can recognize many different epitopes

d. macrophages differentiate into T cells

6. Which of the following is NOT an antibody function?

a. presenting antigen to T Helper cells

b. opsonization

c. neutralizing toxins

d. fixing complement

7. If a dendritic cell presents antigen on MHC I, a ____ will _____.

a. Helper T cell:boost a B cell antibody response

b. Cytotoxic T cell: kill nearby B cells

c. Helper T cells:boost the dendritic cells' antibody response

d. Cytotoxic T cell: kill the dendritic cell

8. Which of the following is NOT a B cell activation response?

a. differentiate to plasma cells

b. clonal expansion

c. producing antibody

d. kill the antigen presenting cell

9. The MHC class of genes that is found on all nucleated cells and is important in tissue rejection of transplants is ___.

a. MHC III

b. MHC II

c. MHC I

d. TNF gene group

10. If you had been bitten by a lethal snake and the doctor had an active or a passive immunization procedure, which would you want and why?
    a. active, so immunity would last
    b. passive, its quicker
    c. active, its quicker
    d. passive, so immunity would last

11. An anamnestic response :
    a. lasts longer than a primary response
    b. is stronger than a secondary response
    c. produces only monoclonal antibodies
    d. is primary IgM

12. If a B cell encounters antigen and presents it to a T helper cell, the antigen will be presented on a _____ molecule.
    a. MHC III
    b. antibody
    c. MHC II
    d. MHC I

13. B cells are __ while T cells are ____.
    a. circulating: located in the thymus
    b. located in the bone marrow only: circulating
    c. circulating :non-circulating
    d. non-circulating :circulating

14. Immunoglobulins are also called ____ and are produced by ___.
    a. lymphokines: T cells
    b. cytokines:B cells
    c. antigens: macrophages
    d. antibodies : B cells

15. If a person had high levels of IgE circulating in his or her bloodstream, this person is likely ___.
    a. fighting a viral infection
    b. in the middle of an allergic response
    c. in the middle of a primary antigen response
    d. in the middle of a secondary antigen response

Chapter 16 Immunization and Immune Assays

Building Your Knowledge

1) In what two separate ways can we apply our knowledge of the immune system to help in the fight against infectious disease?

    a. _____

    b. _____ .

2) How does immunotherapy differ from active immunization?

3) What is ISG therapy?

How is it administered?

Who is it administered to?

How long does the protection last?

4) How does SIG therapy differ from ISG therapy?

What are the advantages of SIG therapy?

What are the drawbacks?

5) What are 6 characteristics of an ideal vaccine?

a. _____

b. _____

c. _____

d. _____

e. _____

f. _____

6) Give an example of a killed or inactivated vaccine.

How are such vaccines prepared?

What is one drawback of killed vaccines over live vaccines?

7) Give an example of an attenuated vaccine.

What are 3 advantages of live vaccines?

What are 3 disadvantages of live vaccines?

8) Which type of vaccines are toxoids (attenuated or inactivated) ?

How are toxoids made?

What do they protect against?

9) List three diseases that there is no reliable vaccine for?

      a. _____

      b. _____

      c. _____

10) What are "Trojan Horse" vaccines and how are they constructed?

What carriers have been used?

What vaccines have been developed and used experimentally?

11) How are DNA vaccines different from attenuated or inactivated vaccines?

12) Why is there an increased interest in smallpox, anthrax, botulism and plague vaccines?

13) Why are oral vaccines better than shots? (list 3 reasons)

      a. _____

      b. _____

      c. _____

14) What are adjuvants and what are they used for?

List 3 commonly used adjuvants.

   a. _____

   b. _____

   c. _____

15) What are the most common complications from vaccination?

16) What is herd immunity?

What does it prevent?

17) What is serology?

18) Differentiate between specificity and sensitivity in serological testing.

19) List four ways antigen-antibody reactions may be detected.

   a. _____

   b. _____

   c. _____

   d. _____

20) Differentiate between agglutination and precipitation.

How do the antigens in each test differ?

How are the processes similar?

21) Give 3 examples of specific agglutination and precipitation tests.

Agglutination tests	Precipitation tests

22) Diagram the steps of a Western Blot. What information does a Western blot tell you?

23) Diagram a simple complement fixation test. What 4 components are needed to complete this test?

24) Differentiate between direct and indirect testing methods using fluorescent antibodies.

Direct –

Examples -

Indirect -

Examples –

25) Compare and contrast capture and direct ELISA methods. What do each detect?

26) Diagram the process of an indirect ELISA.

27) What is a tuberculin test?

How is it performed?

What does a positive test indicate?

Organizing Your Knowledge

Vaccine Type	Advantage	Drawback	Example
Live, Attenuated			
Killed/Inactivated			
Subunit			
Recombinant			

Test Name	Type of Test	Used to detect
RBC clumping		
Widal Test		
Rapid Plasmin Reagin		
Cold Agglutination Test		
Viral Hemagglutination		
Oucherlony		
VDRL		
Immunoelectrophoresis		
Anti-Streptolysin O		
Quellung Test		
RIST		
ELISA		

Fun with Your Knowledge

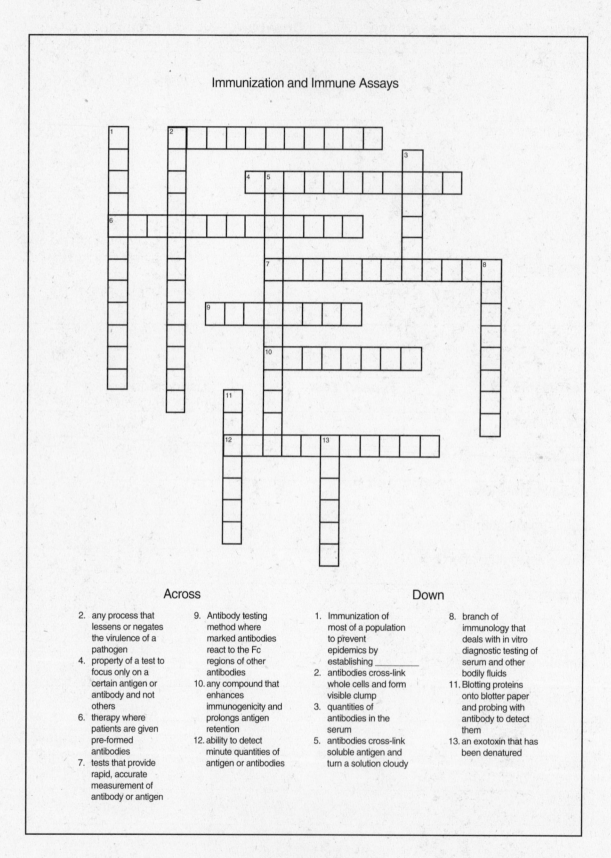

Immunization and Immune Assays

Across

2. any process that lessens or negates the virulence of a pathogen
4. property of a test to focus only on a certain antigen or antibody and not others
6. therapy where patients are given pre-formed antibodies
7. tests that provide rapid, accurate measurement of antibody or antigen
9. Antibody testing method where marked antibodies react to the Fc regions of other antibodies
10. any compound that enhances immunogenicity and prolongs antigen retention
12. ability to detect minute quantities of antigen or antibodies

Down

1. Immunization of most of a population to prevent epidemics by establishing _____
2. antibodies cross-link whole cells and form visible clump
3. quantities of antibodies in the serum
5. antibodies cross-link soluble antigen and turn a solution cloudy
8. branch of immunology that deals with in vitro diagnostic testing of serum and other bodily fluids
11. Blotting proteins onto blotter paper and probing with antibody to detect them
13. an exotoxin that has been denatured

Practicing Your Knowledge

1. Vaccination strategies that are currently being tested include _____.
    a. DNA vaccines
    b. subunit vaccines
    c. Trojan horse vaccines
    d. All of the above

2. The branch of immunology that includes the in vitro diagnostic testing of serum.
    a. sensitivity
    b. serology
    c. specificity
    d. precipitation

3. One advantage of passive immunization, such as receiving ISG or SIG is _____.
    a. the length of immunity
    b. the production of memory cells
    c. the ease of delivery (oral dosing common)
    d. the speed at which immunity is conferred

4. The tuberculin test is a form of ___ testing.
    a. in vitro
    b. ELISA
    c. RAST
    d. in vivo

5. Attenuated vaccines are _____.
    a. live vaccines
    b. heat killed organisms
    c. bits of immunogenic DNA
    d. made up of antigenic subunits of a pathogenic organism

6. The complement fixation test detects the presence of _____ in serum.
    a. toxins
    b. antigen
    c. antibody
    d. bacterial cells

7. The RIST test will measure the level of IgE circulating in allergic patients.
    a. True
    b. False

8. Which of the following items would NOT be the target of an agglutination test?
    a. red blood cells
    b. bacterial cells
    c. exotoxins
    d. latex beads

9. In order to be effective a vaccine must be _____.
    a. antigenic but not immunogenic
    b. immunogenic but not antigenic
    c. pathogenic but not immunogenic
    d. antigenic but not pathogenic

10. Which type of white blood cell readily forms rosettes when mixed with sheep red blood cells?
    a. neutrophils
    b. T cells
    c. B cells
    d. macrophages

11. A immunofluorescence test using antibodies in solution to bind to target cells is a(n) _____ test.
    a. Direct
    b. Western Blot
    c. Indirect
    d. Immunoelectrophoresis

12. If a test can detect very small amounts of an antigen, such as tetanus toxin, but also detects botulism toxin, we can say that it has _____.

    a. sensitivity without specificity
    b. titers without sensitivity
    c. specificity without sensitivity
    d. specificity without titers

13. The goal of immunization is to establish _____ and prevent _____.

    a. individual immunity : polarization
    b. herd immunity : epidemics
    c. adjuvantation:epidemics
    d. indirect immunity : meningitis

14. Vaccines have been developed against all of the following diseases EXCEPT ___.

    a. measles
    b. tetanus
    c. HIV
    d. chicken pox

15. A double-diffusion or Ouchterlony test is a type of _____ test.

    a. agglutination
    b. vaccination
    c. ELISA
    d. precipitation

Chapter 17 Disorders in Immunity

Building Your Knowledge

1) In what two generalized ways can the immune system malfunction and cause disease?

    a. _____

    b. _____

2) Compare and contrast the four separate types of hypersensitivity. How are all 4 types similar? In what ways are they different?

3) Where do exogenous antigens come from?

Where do endogenous antigens come from?

4) Why is it easy to mistake an allergic reaction for an infection?

5) Approximately how many people in the US suffer from hay fever and asthma each year?

How significant are these numbers (what % of the US population)?

6) How can a child inherit an allergy to pollen from a parent who is allergic to dust mites?

7) What is the cause of a type I allergic responses?

What symptoms are commonly seen in allergic individuals?

8) List and give examples of 4 separate portals of entry and allergens.

Entry Portal	Allergen

9) How can allergic individuals not have symptoms of allergy on their first exposure to an allergen, such as bee venom, but nearly die from anaphylaxis on their second exposure?

10) Which antibody class is associated with type I allergies?

Which cell type do these antibodies bind to?

11) How are asthma, eczema, food and drug allergies similar?

What are the symptoms of each allergy?

asthma -

eczema -

food allergies –

12) What is the difference between systemic and cutaneous anaphylaxis?

Which can kill in 15 minutes?

13) How can allergies be diagnosed?

What chemicals or molecules are indicative of an active immune response?

14) What three methods are there to treat or prevent allergies attacks?

    a. _____

    b. _____

    c. _____

15) How does desensitization to allergens prevent allergy attacks?

16) How do type II hypersensitivities differ from type I hypersensitivities?

Which cells mediate type II hypersensitivities?

What is the target of type I hypersensitivities?

17) What are the ABO antigens?

If a person has an allele for type O and one for type A, what will that person's blood type be?

Which blood type is the universal donor? Why?

Which blood type is the universal recipient? Why?

18) What would happen if type B blood was given to someone with type A blood?

How is a transfusion reaction treated?

19) What do Rh+ and Rh- mean?

How are Rh- females commonly sensitized to Rh+ antigens?

What is erythroblastosis fetalis and how is it caused?

How does RhOGAM prevent hemolytic disease of the newborn?

20) What are type III hypersensitivities caused by? How are type II and type III sensitivities similar and how are they different??

21) What are the symptoms of serum sickness?

What are these symptoms caused by?

22) What is the Arthus reaction?

How are serum sickness and the Arthus reaction similar?

How are they different?

23) Which type(s) of hypersensitivity are responsible for most autoimmune reactions?

24) Are males or females more commonly diagnosed with autoimmune disease?

25) Compare and contrast the four theories of auto-immune reaction development.

26) How are SLE and rheumatoid arthritis similar?

How are they different?

27) Name and describe an endocrine and a neuromuscular autoimmune diseases.

28) Which cells mediate type IV hypersensitivities?

29) Which class of TH cells cause the tuberculin reaction?

    What does a positive tuberculin test indicate?

30) Why does poison ivy exposure cause itching?

31) What immune cells and molecules (markers) are responsible for transplant rejection?

32) What cells are transplanted to cause GVHD?

    How is GVHD similar to rejection of foreign tissue?

    How is GVHD treated?

33) Where may donor organs and tissues come from?

    Which types are more successful?

34) What drugs are used to limit tissue rejection and how do they work?

35) Differentiate between primary and secondary immunodeficiency diseases.

Primary -

Secondary -

36) What diseases are associated with a lack of or deficiency in antibody production?

What infections are these patients prone to get?

37) What is DiGeorge syndrome and what problems are associated with it?

38) What is the "boy in the bubble" syndrome? What defects may cause SCID?

39) What 4 general ways can a person acquire an immunodeficiency?

a. _____

b. _____

c. _____

d. _____

40) What exactly is cancer?

What is the difference between a benign and a malignant tumor?

What is metastasis?

41) What are the most common cancers in the US? In men?   In women?

How many new cases of cancer are diagnosed each year?

How many people die?

42) How do genetics influence cancer development and progression?

43) How are viruses and cancer linked?

44) What chromosomal alterations are associated with cancer?

45) Why are AIDS patients more prone to certain type of cancer?

Organizing Your Knowledge

Immunopathology	Over-Activity?	Under-Activity?
Allergies		
Graft Rejection		
Immunodeficiency		
Autoimmunity		

Theory	Reason for Autoimmune Reaction
Sequestered Antigen	
Immune deficiency	
Clonal Selection	
Molecular Mimicry	

Hypersensitivity Reaction	Cells or Antibodies Involved	Type of Hypersensitivity
Asthma		
Rheumatoid Arthritis		
Contact Dermatitis		
Pernicious Anemia		
Hay fever		
Graft Rejection		
Serum Sickness		
Anaphylaxis		
Eczema		
Arthus reaction		

Immunodeficiency	Immune component Affected	Primary or secondary?
Pregnancy		
Transplant drugs		
ADA deficiency		
Chronic Granulomatous disease		
DiGeorge Syndrome		
Stress		
Steroids		
Hypogammaglobulinemia		
AIDS		

Allergic Mediator	Actions on the Body	Symptoms
histamine		
serotonin		
leukotrienes		
bradykinin		
prostaglandins		
platelet activating factor		

Graft type	Donor	Example
Xenograft		
Isograft		
Allograft		
Autograft		

Drug or Therapy	Allergy Mediator Blocked
corticosteroids	
cromolyn	
Xolair	
theophylline	
epinephrine	
antihistamines	
desensitization	

Causative Agent	Specific Example
oncogenes	
viral infection	
chromosomal damage	
programmed cell death errors	
carcinogen exposure	
immunosuppression	

Practicing Your Knowledge

1. A person with hay fever may pass on a mold allergy to his or her children.
    a. True
    b. False

2. Which of the following is NOT typically an inhaled allergen?
    a. pollen
    b. animal hair
    c. antibiotics
    d. mold spores

3. Immediate hypersensitivities are caused by _____ reactions.
    a. IgE
    b. neutrophil
    c. macrophages
    d. IgM

4. Which of the following is NOT an autoimmune disease?
    a. Rheumatoid arthritis
    b. DiGeorge syndrome
    c. Diabetes mellitus
    d. Multiple sclerosis

5. Antigens that arise from self tissue and cause a hypersensitivity are ___ and cause _____ reactions.
    a. exogenous:IgE
    b. endogenous:IgE
    c. exogenous:autoimmune
    d. endogenous:autoimmune

6. A person with type A blood will have antibodies against _____.
    a. type B blood
    b. type O blood
    c. type A blood
    d. none of the above - type A is the universal donor

7. The sequestered antigen theory explains the development of ___.
    a. cancer
    b. acquired immunodeficiency
    c. autoimmune disease
    d. allergic hypersensitivities

8. Which of the following are type I hypersensitivities?
   a. food allergies
   b. asthma
   c. eczema
   d. all of the above

9. It is often difficult to distinguish and allergic response from an active infection because _____.
   a. both are contagious
   b. both are inflammatory responses
   c. both cause cancer
   d. both are mediated by red blood cells

10. A transplant from one species to another, such as the use of pig heart valves in humans, is called a ___.
   a. allograft
   b. xenograft
   c. isograft
   d. autograft

11. RhoGAM is used to treat ___ mothers of ___ children, to prevent damage to future offspring.
   a. Rh+ , Rh-
   b. Rh-, Rh+
   c. AB+, O-
   d. A, B

12. Which of the following is NOT a mediator of type I hypersensitivity?
   a. antigen-antibody complexes
   b. IgE
   c. histamine
   d. bradykinin

13. Serum sickness and the Arthus reaction are both ___.
   a. systemic diseases
   b. type I hypersensitivities
   c. autoimmune reactions
   d. type III hypersensitivities

14. A primary immunodeficiency ___.
   a. is acquired through repeated exposure to allergens
   b. is present from birth and may lead to anaphylaxis
   c. is present from birth and may lead to opportunistic infections
   d. is acquired through exposure to immunosuppressive drugs, such as steroids

15. How may allergies be treated or prevented?
   a. hypersensitization
   b. taking histamines
   c. desensitization therapy
   d. tissue transplantation

Chapter 18 The Cocci of Medical Importance

Building Your Knowledge

1) What are the pyogenic cocci? (List the 4 genera discussed in your book)

   a. _____

   b. _____

   c. _____

   d. _____

Which are Gram positive? Which are Gram negative?

2) What are the general characteristics of *Staphylococci*?

What do the staphylococci look like?

Where are they found?

Do *Staphylococci* form spores?

Are they generally flagellated or encapsulated?

Which staphylococcus is considered the most serious human pathogen from the group?

3) How well does *S. aureus* handle environmental stress?

Why is this significant?

4) List four separate enzymes produced by *S. aureus* and how these enzymes contribute to the disease process.

5) What specific toxins are associated with S. aureus and how do they cause damage?

6) How common is staphylococcal infection or colonization?

7) What is MRSA and how is it spread?

8) Differentiate between furuncles, carbuncles and impetigo. How are they similar? How are they different?

9) What traits of S. aureus make it a leading cause of food poisoning?

10) Systemic staphylococcal infection can take many forms. Describe 3.

11) How are TSS and SSSS similar?

    How are they different?

12) Are antibodies, cell-mediated immunity or phagocytes most protective against S.
    aureus infection?

13) Name 2 coagulase negative staphylococci and the diseases they cause.

14) What test can differentiate staphylococci from streptococci?

15) What is the easiest way to distinguish *S. aureus* from the other staphylococci of
    medical concern?

16) Differentiate between coagulase and catalase enzymes. What is the function of
    each?

17) How are staphylococcal infections treated?

    Is penicillin the drug of choice?   Why or why not?

18) How may the spread of *S. aureus* be limited?

19) What are the general characteristics of streptococci? (appearance, diseases, growth,etc)

Do streptococci grow as easily as staphylococci in lab?

20) How are streptococci classified? (Describe 2 methods)

a. _____

b. _____

21) Differentiate between alpha-hemolysis and beta-hemolysis.

22) What are the streptococci most commonly associated with human disease?

23) How do M proteins and capsules contribute to streptococcal pathogenesis?

24) What toxins and enzymes are produced by streptococci?

25) How does streptococcal infection spread?

What types of infections are seen most in the summer?

In winter?

26) Describe and differentiate between pyoderma, streptococcal pharyngitis, and scarlet fever.

27) What are the potential long-term complications of strep throat?

28) What antibodies give long-term protection against group A streptococcal infection?

29) What are the group B streptococci?

Where are they found?

What populations are most at risk for group B streptococcal infection?

How are infections controlled or prevented?

30) What are the enterococci?

Where are they found?

Why are they of concern to human health?

31) Why are rapid strep tests so valuable?

32) How are rheumatic fever and glomerulonephritis treated?

33) What are viridans streptococci and what problems do they cause?

34) Why is *S. pneumoniae* also called pneumococcus?

35) Are pathogenic strains of *S. pneumoniae* smooth or rough? Why?

36) How is pneumococcus spread?

   Are fomites a primary source of infection? Explain.

37) What 2 diseases are most commonly associated with pneumococcal infection?

   How are pneumococcal infections prevented and treated?

38) Draw and indicate the color of Neisseria species.

39) How are Neisseria grown in lab? Are they easy or difficult to culture?

40) What species are infected with *N. gonorrhoeae*?

   How is the bacterial disease spread?

   What are the symptoms of infection?

   What areas of the body are commonly affected?

41) Differentiate between meningococcus, gonococcus, and pneumococcus.

42) Who is most at risk for meningococcal disease?

   How is meningococcal disease spread?

   What are the symptoms of infection with meningococcus?

   How is meningococcal infection treated?

43) List three other cocci of medical importance and the diseases they cause.

Organizing Your Knowledge

Protein produced by *S. aureus*	Action in the body	How this aids the bacteria
coagulase		
hyaluronidase		
staphylokinase		
penicillinase		
leukocidin		
TSST		

Diseases	Areas Affected	Symptoms
folliculitis		
osteomyelitis		
bacteremia		
food intoxication		
scalded skin syndrome		

Organism	S. aureus	S. pneumoniae	S. pyogenes	N. gonorrhoeae	N. meningitidis
Appearance **(Draw 5 cells)**					
Gram stain					
Disease caused					
Diagnostic test					
Virulence traits					
Treatment					

Practicing Your Knowledge

1. Group B streptococcal infections often cause serious problems in ___.
    a. college students
    b. adult females
    c. newborns
    d. sexually active males

2. Gonococcus is a ____ that causes ____.
    a. gram negative coccus: ear infections
    b. gram positive coccus: pneumonia
    c. gram negative coccus: gonorrhea
    d. gram positive coccus: gonorrhea

3. The most powerful defense humans have against *S. aureus* is ___.
   a. Antibody production by B cells
   b. Cytotoxic T cell killing
   c. Neutrophils and Macrophages
   d. Mast cells and Eosinophils

4. Gonococcus, meningococcus, and pneumococcus are all ___.
   a. causative agents of strep throat
   b. gram negative
   c. gram positive
   d. cocci

5. Gonococcal infections can cause blindness in infants.
   a. True
   b. False

6. Penicillin is the drug of choice for treating *S. aureus*.
   a. True
   b. False

7. Smooth strains of *S. pneumonia* _____ and are ___.
   a. have pili : virulent
   b. lack a capule : avirulent
   c. have a capsule : virulent
   d. lack pili: avirulent

8. Which of the following characteristics is typical of *S. aureus*?
   a. low salt tolerance
   b. coagulase negative
   c. spore-forming
   d. toxin-producing

9. Pneumococcus commonly causes ____ and pneumonia.
   a. endocarditis
   b. otitis media
   c. dental caries
   d. kidney infections

10. The causative agent of "strep throat" is a ___.
   a. streptococci
   b. staphylococci
   c. meningococci
   d. streptobacilli

Chapter 19 The Gram-Positive Bacilli of Medical Importance

Building Your Knowledge

1) What is an endospore?

   From the bacterial standpoint, what is the advantage to producing endospores?

   Do ALL Gram-positive rods form endospores? Explain.

2) List the 3 Gram positive genera that are endospores-formers.

    a. _____

    b. _____

    c. _____

3) What are the symptoms of anthrax?

In which regions, both worldwide and in the US, is anthrax commonly found?

What microbe causes anthrax and where does that microbe commonly live?

How do people in the United States commonly acquire anthrax?

What forms may an anthrax infection be found in? Which is the most serious?

How is anthrax treated and controlled?

4) How can you differentiate between *Bacillus* and *Clostridium*?

What is the causative agent of gas gangrene?

Where is this bacterium commonly found?

How is gas gangrene treated and prevented?

Which is more serious, anaerobic cellulitis or myonecrosis? Why?

Why does hyperbaric oxygen therapy effectively treat *Clostridium*?

Why doesn't immunization work to prevent gas gangrene?

5) What disease is caused by *Clostridium difficile*? _____

   Who is at risk to develop this disease?

   How is the disease treated or prevented?

6) What bacterial species causes tetanus? _____

   Do these bacteria form endospores? Why is this important?

   Are scrapes or puncture wounds more likely to become infected with the tetanus bacilli? Why?

   What are the symptoms of tetanus?

   What is the mortality rate for tetanus? _____

   How is tetanus prevented and treated?

7) What microbe is the causative agent of botulism? _____

   Is botulism an infection or an intoxication? Explain.

   How is infant botulism contracted?

   How is botulism prevented and treated?

8) How can clinical labs differentiate between the different clostridial species?

9) What are the pathogenic regular genera of non-spore-forming Gram positive bacilli?

     a. _____

     b. _____

10) What is listeriosis and how is it contracted?

    What microbe causes listeriosis? _____

    Who is most at risk to develop listeriosis?

    How is listeriosis spread?

    How is listeriosis treated?

11) What disease does *Erysipelothrix rhusiopathiae* cause? _____

    Who is most at risk to contract the disease?

    Which species generally contract erysipeloid?

    How is the disease treated?

12) What are the genera of the pathogenic non-spore-forming gram-positive bacilli?

     a. _____

     b. _____

     c. _____

13) What is the causative agent of diphtheria? _____

How is diphtheria transmitted?

Who is at risk for contracting diphtheria?

What is diphtherotoxin and how does it contribute to the disease process?

How is diphtheria diagnosed and treated?

14) What bacterial species most associated with the development of acne?

_____

15) What are the distinguishing characteristics of the mycobacteria?

Do mycobacteria produce spores?

Where do most mycobacteria live in the environment?

Why are mycobacteria acid-fast?

16) What is the causative agent of tuberculosis? _____

Is tuberculosis an emerging infectious disease or ancient disease? Explain.

What is the infectious dose of *M. tuberculosis*?

Do all infected individuals become ill?

Why is tuberculosis on the rise?

What is DOT and why is it important in the treatment of tuberculosis?

17) Where can infection with tuberculosis cause symptoms?

What is a tubercle and how is it formed?

How is tuberculosis diagnosed?

What is BCG and when is it given?

18) What is the causative agent of leprosy? _____

Where is leprosy endemic?

What animals are used to study leprosy infection?

What are the major differences between tuberculoid and lepromatous leprosy?

How is leprosy treated?

How long does the treatment course last?

19) Who is most susceptible to non-tuberculous mycobacterial infection? Why?

What are the non-tuberculus mycobacteria (NTM) and what diseases do they cause?

20) What diseases are caused by pathogenic actinomycetes?

21) What diseases are caused by pathogenic *Nocardia* ?

Organizing Your Knowledge

Organism	Shape	Differential Stain Traits	Spore-former
Bacillus anthracis			
Bacillus cereus			
Clostridium perfringens			
Clostridium difficile			
Clostridium tetani			
Clostridium botulinum			
Mycobacterium tuberculosis			
Mycobacterium leprae			
Listeria monocytogenes			
Erysiplothrix rhusiopathiae			
Proprionibacterium acnes			
Corynebacterium diphtheriae			
Actinomyces			
Nocardia			

Organism	Disease(s) Caused	Populations at Risk	Treatment	Prevention
Bacillus anthracis				
Bacillus cereus				
Clostridium perfringens				
Clostridium difficile				
Clostridium tetani				
Clostridium botulinum				
Mycobacterium tuberculosis				
Mycobacterium leprae				
Listeria monocytogenes				
Erysiplothrix rhusiopathiae				
Proprionibacterium acnes				
Corynebacterium diphtheriae				
Actinomyces				
Nocardia				

Fun with Your Knowledge

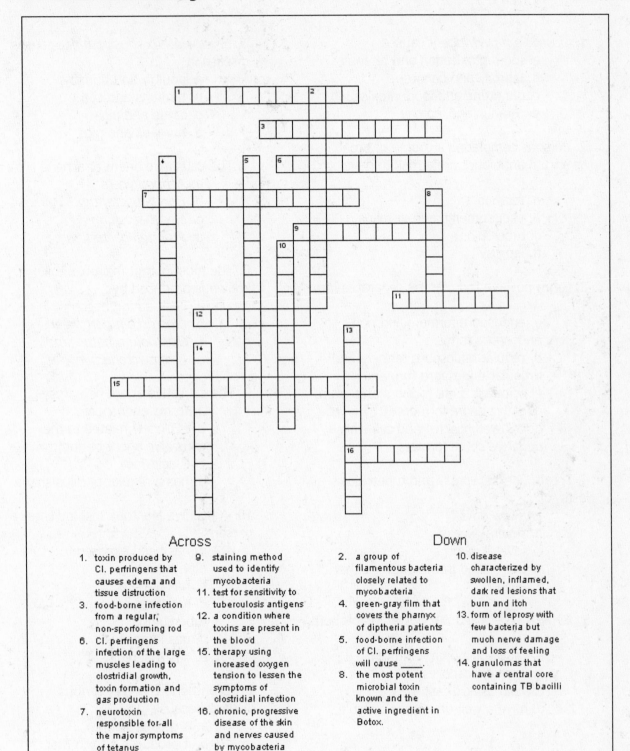

Across

1. toxin produced by Cl. perfringens that causes edema and tissue distruction
3. food-borne infection from a regular, non-sporforming rod
6. Cl. perfringens infection of the large muscles leading to clostridial growth, toxin formation and gas production
7. neurotoxin responsible for all the major symptoms of tetanus
9. staining method used to identify mycobacteria
11. test for sensitivity to tuberculosis antigens
12. a condition where toxins are present in the blood
15. therapy using increased oxygen tension to lessen the symptoms of clostridial infection
16. chronic, progressive disease of the skin and nerves caused by mycobacteria

Down

2. a group of filamentous bacteria closely related to mycobacteria
4. green-gray film that covers the pharnyx of diptheria patients
5. food-borne infection of Cl. perfringens will cause ____.
8. the most potent microbial toxin known and the active ingredient in Botox.
10. disease characterized by swollen, inflamed, dark red lesions that burn and itch
13. form of leprosy with few bacteria but much nerve damage and loss of feeling
14. granulomas that have a central core containing TB bacilli

Practicing Your Knowledge

1. *Clostridium perfringens* causes ___.
    a. food intoxication and tetanus
    b. leprosy and gangrene
    c. gangrene and food intoxication
    d. tetanus and leprosy

2. Persons completing a course of broad-spectrum antibiotics are at risk to develop ___.
    a. tetanus
    b. pseudomembranous colitis
    c. tuberculosis
    d. leprosy

3. Gram-positive rods can be separated into __
    a. endospore formers and non-endospore formers
    b. regular endospore formers and irregular endospore formers
    c. acid fast spore formers and gram negative spore formers
    d. cell wall positive and cell wall negative spore formers

4. Treatment for leprosy and tuberculosis lasts for ___.
    a. a few days
    b. several weeks
    c. several months
    d. there is no treatment for either disease

5. *Bacillus cereus* and *Bacillus anthracis* both __.
    a. cause pulmonary anthrax
    b. form endospores
    c. stain as acid-fast rods
    d. have a well-developed capsule

6. *Erysipelothrix rhusiopathiae* causes disease in ___.
    a. poultry and humans
    b. humans and cattle
    c. cattle and pigs
    d. humans and pigs

7. The causative agent of acne is a(n):
    a. *Actinomycete*
    b. *Corynebacterium*
    c. *Mycobacterium*
    d. *Proprionibacterium*

8. The most potent microbial toxin known is produced by ___.
    a. flesh-eating bacteria
    b. *Clostridium perfringens*
    c. *Clostridium botulinum*
    d. *Mycobacterium leprae*

9. The causative agent of leprosy ___.
    a. forms endospores.
    b. is closely related to the causative agent of anthrax
    c. is acid-fast
    d. has a regular bacillus shape

10. A positive Mantoux test indicates a person ___.
    a. has been exposed to diphtheria
    b. has tetanus
    c. has been exposed to tuberculosis
    d. has leprosy
    e. has tuberculosis

11. Diphtheria is a very serious disease because ___.
    a. it causes lockjaw
    b. it produces a toxin that stops protein synthesis
    c. it produces a neurotoxin
    d. it converts into a deadly virus once inside a host

12. Which of the following infections is routinely prevented by vaccination?
    a. leprosy
    b. tuberculosis
    c. gas gangrene
    d. tetanus

13. The causative agents of tetanus and botulism are both ___.
    a. obligate anaerobes
    b. obligate aerobes
    c. mycobacteria
    d. food-borne pathogens

14. The causative agents of tuberculosis and tetanus both:
    a. are acid-fast
    b. form endospores
    c. are obligate anaerobes
    d. are bacilli
    e. all of the above

15. Listeriosis is a(n) _____.
    a. form of tuberculosis
    b. food-borne infection
    c. type of wound infection
    d. intoxication, not infection

Chapter 20 Gram-Negative Bacilli of Medical Importance

Building Your Knowledge

1) What three categories can be used to separate ALL Gram-negative bacilli?

    a. _____

    b. _____

    c. _____

2) Where are enteric bacteria found?

3) Why is lipopolysaccharide (LPS) dangerous?

    Where is LPS found?

4) Where is *Pseudomonas aeruginosa* normally found in the environment?

What diseases does it cause?

Who is most at risk to develop a pseudomonas infection? (give several groups)

A vaccine for *P. aeruginosa* was recently tested in cystic fibrosis patients. Why are these patients good candidates for this vaccine?

5) Who is at risk for *Burkholderia cepacia* infection?

6) What are the common names for brucellosis?

How are pigs and cows affected by brucellosis?

7) How are humans affected by brucellosis?

Have vaccines been effective against brucellosis in the past?

How is brucellosis diagnosed and treated?

8) What is rabbit fever caused by?

Where is tularemia endemic?

What are the major symptoms and treatment for tularemia?

Why has there been an increase in interest for tularemia research in the past few years?

9) What is the causative agent of whooping cough?

How does infection lead to the "whoop" sound associated with whooping cough?

Why is whooping cough on the rise?

How is whooping cough diagnosed and treated?

10) How did *Legionella* get its name?

Where is *Legionella* commonly found in the environment?

How is Legionnaires disease diagnosed and treated?

11) What group of pathogens commonly causes diarrheal disease?

How many diarrheal infections occur are estimated to occur each year ?

Why is this just an estimate?

12) What is the difference between coliforms and non-coliforms?

How can you distinguish an *E. coli* from a *Salmonella*?

13) What is enrichment media and why is it used?

14) Draw a Gram-negative enteric rod, labeling the K, O and H antigens. Give the full names for the structures corresponding to each antigen (e.g. K= capsular antigen)

Why is serotyping enteric bacteria important to epidemiologists?

15) Why is *E. coli* sometimes considered the predominant bacterial species in the human intestine?

16) How do enterotoxigenic, enteroinvasive and enteropathogenic *E. coli* differ from one another?

What other infections is each similar to?

17) What is the single greatest cause of death among babies (worldwide)?

What is the most common cause of "traveler's diarrhea"?

Why is it better to use Pepto-Bismol than kaolin to treat traveler's diarrhea?

What is the most common source of urinary tract infections (UTIs)?

Why are UTIs more commonly seen in women?

18) Are the other coliforms generally true pathogens or opportunists?

Why are they of clinical concern?

19) What diseases do *Klebsiella, Enterobacter, Serratia and Citrobacter* cause?

20) How are Salmonella and Shigella different from the coliforms?

21) What is the causative agent for typhoid fever?

Who is the natural host for the pathogen?

How is typhoid fever transmitted from host to host?

What are the major symptoms of typhoid and how is it treated?

22) How do enteric fevers differ from typhoid fever?

a. _____

b. _____

c. _____

What is the current estimate of the number of non-typhoidal salmonellosis cases each year?

Which animals are particularly associated with Salmonella infection?

How is Salmonella infection diagnosed and treated?

23) What disease does Shigella cause?

What is the natural host for Shigella?

How does Shigella infection differ from Salmonella infection?

What is the treatment for shigellosis?

24) What are the enteric *Yersinia*?

What disease do they cause?

25) What is the causative agent of plague?

What are the virulence factors associated with plague? (list 3)

Where is plague endemic?

26) Are humans endemic reservoirs for plague? Explain.

27) Differentiate between sylvanic, urban and pneumonic plague.

How is each transmitted?

Why is plague also called "the Black Death"?

28) How is plague treated?

What is the survival rate, with treatment?

29) List 3 species of *Haemophilus* and the diseases they cause.

Haemophilus	Disease Caused

Organizing Your Knowledge

Organism	Gram Stain & Shape	Oxygen Requirement	Diseases caused
Bordetella pertussis			
Brucella abortus			
Burkolderia cepia			
Citrobacter			
Enterobacter			
Escherichia coli			
Francisella tularensis			
Haemophilus aegyptius			
Haemophilus influenzae			
Klebsiella pneumoniae			
Legionella pneumophila			
Pseudomonas aeruginosa			
Salmonella			
Serratia			
Shigella			
Yersinia enterocolitica			
Yersinia pestis			

Organism & Disease	Reservoir & Transmission	At risk populations	Diagnostic tests	Treatment or Prevention
Bordetella pertussis				
Brucella abortus				
Burkolderia cepia				
Citrobacter				
Enterobacter				
Escherichia coli				
Francisella tularensis				
Haemophilus aegyptius				
Haemophilus influenzae				
Klebsiella pneumoniae				
Legionella pneumophila				
Pseudomonas aeruginosa				
Salmonella				
Serratia				
Shigella				
Yersinia enterocolitica				
Yersinia pestis				

Practicing Your Knowledge

1. The DTaP vaccine protects against:
   a. diphtheria, polio, tetanus
   b. polio, tuberculosis, diphtheria
   c. diphtheria, tetanus, pertussis
   d. diphtheria, tuberculosis, pertussis

2. Legionnaire's pneumonia and Pontiac fever are both ____.
   a. respiratory illnesses
   b. caused by enteric pathogens
   c. diarrheal illnesses
   d. caused by *Legionella*

3. The causative agents of diphtheria and whooping cough are both:
   a. Gram-positive
   b. Gram-negative
   c. bacilli
   d. spore-formers

4. Which of the following is NOT an enteric pathogen?
   a. *E. coli*
   b. *Salmonella*
   c. *Shigella*
   d. *Haemophilus*

5. Vaccines to protect against *Pseudomonas* infection have been tried in _____ and were ____.
   a. pregnant women : unsuccessful
   b. cattle farmers : successful
   c. cystic fibrosis patients : successful
   d. burn victims : unsuccessful

6. *Serratia*, *Citrobacter* and *Enterobacter* are all ____.
   a. true pathogens
   b. obligate aerobes
   c. opportunistic coliforms
   d. non-enteric, non-coliforms

7. Which of the following is the causative agent of the plague?
   a. *Yersinia pestis*
   b. *Shigella flexni*
   c. *Klebsiella pestis*
   d. *Yersinia enterocolitica*

8. Enrichment media is used to ____.
   a. inhibit pathogen growth
   b. favor normal flora growth
   c. inhibit normal flora growth
   d. all of the above

9. *Pseudomonas aeruginosa* is a particular threat to ____.
   a. rabbit hunters
   b. burn victims
   c. pregnant women
   d. professional painters

10. The causative agent of "flu" is ____.
   a. *Yersinia pestis*
   b. a virus
   c. *Pasteurella multocida*
   d. *Haemophilus influenzae*

11. *Salmonella* and *Shigella* are both :
   a. causative agents of typhoid fever
   b. true pathogens
   c. coliforms
   d. Gram positive bacilli

12. Enteric pathogens most often cause __ diseases.
   a. respiratory
   b. diarrheal
   c. skin
   d. sexually transmitted

13. New cases of pertussis have ___ in the US since 1981.
   a. slowly declined
   b. rapidly declined
   c. rapidly increased
   d. slowly increased

14. Which of the following organisms is NOT a pathogen of concern as a potential bioterrorist threat?
   a. *Francisella tularensis*
   b. *Brucella abortus*
   c. *Bordetella pertussis*
   d. *Yersinia pestis*

15. K antigens are part of the bacterial _____ and are recognized by a host as foreign.
   a. capsule
   b. cell wall
   c. cytoplasm
   d. flagella

Chapter 21 Miscellaneous Bacterial Agents of Disease

Building Your Knowledge

1) Do spirochetes stain gram-negative or gram-positive?

2) How do spirochetes move?

3) Spirochetes are flagellated. Where would you find spirochete flagella?

4) What is the causative agent of syphilis? _____

   How is syphilis transmitted? Why are fomites?

   What is the infectious dose required to develop syphilis?

   Differentiate between primary, secondary and tertiary syphilis?

	Primary	Secondary	Tertiary
Time after exposure			
Major Symptoms			
Treatment options?			

   Why is tertiary syphilis very rare today?

5) How is congenital syphilis contracted and what are the symptoms?

   How is infection with *T. pallidum* diagnosed?

   How is syphilis commonly treated?

6) How are bejel, yaws, and pinta similar to syphilis?   How are these diseases
   different?

7) What is *Leptospira interrogans*?

   What disease does *L. interrogans* cause?

   How is this disease transmitted?

   How is leptospirosis diagnosed, treated?

   What preventative measures can be taken against the disease?

8) How is Borrelia transmitted to humans?

   What is the causative agent of relapsing fever?

   What are the symptoms of relapsing fever?

   How can relapsing fever be prevented or treated?

9) How is Lyme disease transmitted? What are endemic regions for Lyme disease?

What are the early signs of infection with the Lyme disease spirochete?

What organism causes Lyme disease? _____

What are the symptoms of Lyme disease?

How can Lyme disease be prevented or treated?

How is Lyme disease diagnosed?

10) Draw a vibrio.

11) How does someone "catch" cholera?

What are the major symptoms of cholera?

How is cholera spread?

Why is cholera a deadly disease (by what mechanism does it kill) ?

How is cholera diagnosed and treated?

12) What two vibros may be found in seafood?

     a. _____

     b. _____

13) What pathogen is the causative agent of stomach ulcers _____

How is this pathogen transmitted?

How are ulcers now treated and why this course of action?

14) What disease does *Campylobacter jejuni* cause?

How is this disease transmitted?

What are the symptoms and treatment for this infection?

How is *Campylobacter jejuni* infection diagnosed?

What other diseases do *Campylobacter* species cause?

15) What diseases are caused by rickettsias? (list 4)

16) How are rickettsial diseases transmitted?

17) What is the causative agent of typhus? _____

How is typhus transmitted?

What are the major symptoms of typhus?

18) What is Rocky Mountain spotted fever and what causes it?

19) What are the symptoms of Q fever and what is the causative agent of Q fever?

20) Who is at risk to develop cat scratch disease (CSD)?

What is the causative agent of CSD and how is the disease treated?

How can CSD be prevented?

21) Which diseases are caused by *Chlamydia* species?

How are these diseases diagnosed and treated?

Why is azithromycin (an antibiotic that works intracellularly) a good choice to treat *Chlamydia* infections?

22) Which bacteria lack cell walls?

What diseases do they cause?

Why is penicillin a poor choice to treat mycoplasmas and *Chlamydia*?

23) What is the causative agent of Crohn's disease?

24) Draw a tooth, labeling the crown, root, enamel and pulp cavity.

Where is the gingiva?

25) Which oral bacteria are commonly associated with dental caries?

26) Diagram the process of cavity formation from 1st degree caries to 3rd degree caries.

27) Which is more serious, gingivitis or peridontitis and why?

28) How are most dental diseases controlled?

Organizing Your Knowledge

Organism	Describe Organism	Disease(s) Caused	Symptoms
Bartonella			
Borrelia burgdorferi			
Borrelia hermsii			
Campylobacter fetus			
Campylobacter jejuni			
Chlamydia			
Coxiella burnetii			
Ehrlichia			
Helicobacter pylori			
Leptospira interrogans			
Mycoplasma pneumonia			
Rickettsia prowazekii			
Rickettsia Rickettsii			
Treponema pallidum			
Vibrio cholera			
Vibrio parahaemolyticus			

Disease	Reservoir or Source	Mode of Transmission	Diagnosis	Treatment & Prevention
Q Fever				
Syphilis				
Yaws				
Leptospirosis				
Relapsing Fever				
Lyme disease				
Cholera				
Ulcers				
Rocky Mountain Spotted fever				
Cat-Scratch disease				
"walking pneumonia"				
Epidemic typhus				
Ocular trachoma				

Testing Your Knowledge

1. *Erythema migrans* is a characteristic of _____.
    a. relapsing fever
    b. cholera
    c. Lyme disease
    d. tertiary syphilis

2. Campylobacter and cholera are both :
    a. vibrios
    b. arthropod-borne pathogens
    c. respiratory pathogens
    d. gram-positive

3. All spirochetes ___.
   a. cause disease
   b. are sexually transmitted
   c. are acid-fast
   d. are gram-negative

4. Mycoplasmas are __.
   a. all acid fast
   b. always sexually transmitted
   c. lacking cell walls
   d. the causative agents of trench fever

5. Gummas and a hard chancre are both signs of ___.
   a. AIDS
   b. primary syphilis
   c. syphilis
   d. tertiary syphilis

6. Bartonella and Coxiella are both ___.
   a. causes of Q fever
   b. bacterial pathogens
   c. viral pathogens
   d. water-borne pathogens

7. Chlamydial infection can lead to :
   a. blindness
   b. fevers and headache
   c. pelvic inflammatory disease
   d. all of the above

8. Treponemes, rickettsias and chlamydias all ___
   a. require cells for cultivation
   b. are transmitted sexually
   c. are vector-borne pathogens
   d. are energy parasites

9. The natural hosts and source for the bacteria that cause syphilis are ___.
   a. sheep and cows
   b. humans
   c. sheep only
   d. soil

10. The causative agent of walking pneumonia is ___.
    a. vibrio
    b. spirochete
    c. mycoplasma
    d. rickettsia

11. Elementary bodies and reticulate bodies are stages of ____
    a. treponemes
    b. gummas
    c. rickettsias
    d. chlamydia

12. The causative agent of ulcers is ____.
    a. stress
    b. *Helicobacter pylori*
    c. *Vibrio vulnificus*
    d. *Campylobacter jejuni*

13. The life-threatening symptom of cholera infection is ___.
    a. secretory diarrhea
    b. lockjaw
    c. flaccid paralysis
    d. cholera is not a life-threatening disease

14. A disease transmitted by exposure to infected animal urine is ____.
    a. relapsing fever
    b. borreliosis
    c. leptospirosis
    d. yaws

15. Rocky mountain spotted fever is caused by ____.
    a. a chlamydia
    b. a rickettsia
    c. exposure to contaminated water
    d. *Coxiella burnetti*

Chapter 22 The Fungi of Medical Importance

Building Your Knowledge

1) How are humans exposed to fungi?

2) Are most fungi pathogenic to humans? Explain.

3) How do true fungal pathogens differ from opportunistic pathogens? (List 5 separate ways)

   a. _____

   b. _____

   c. _____

   d. _____

   e. _____

4) Which fungal form, yeast or hyphal form, is associated with human infection?

   How does thermal dimorphism contribute to fungal virulence?

5) How are most fungal pathogens transmitted?

   How do fungal epidemics commonly occur? (list 2 specific examples)

6) Name 2 Transmissible fungal infections and indicate how they are spread.

Fungal Agent	Mode of Transmission

7) How do most mycoses agents enter the human body?

8) List four separate fungal virulence traits.

     a. _____

     b. _____

     c. _____

     d. _____

9) Why is culturing fungus not the most common method to diagnose a fungal infection?

   How are fungal infections diagnosed? (list 3 rapid identification methods used)

   Why is identification of a fungal infection (and distinguishing it from a bacterial infection) critical when treating immunocompromised patients?

10) How are fungal infections treated?

11) What 3 basic structures or processes are common targets for anti-fungal drugs?

     a. _____

     b. _____

     c. _____

12) How can fungal infections be prevented?

13) What 2 traits do all primary fungal pathogens share?

     a. _____

     b. _____

14) How is histoplasmosis contracted and who is at risk to contract it?

What are the major symptoms of histoplasmosis?

How does *Histoplasma capsulatum* evade the immune system?

How is histoplasmosis diagnosed and treated?

15) What is Valley Fever?

Where is this pathogen endemic?

What conditions favor the spread of coccidioidomycosis?

What activities are associated with outbreaks of the disease?

What are the major symptoms of coccidioidomycosis?

How is coccidioidomycosis diagnosed and treated?

16) What is the causative agent of blastomycosis?

Where is this disease endemic and how is it transmitted?

What are the symptoms of blastomycosis?

How is blastomycosis diagnosed and treated?

17) What is rose-gardener's disease and what is it caused by?

18) Is sporotrichosis generally a systemic or subcutaneous disease?

Why is pulmonary sporotrichosis on the rise?

19) Are the causative agents of chromoblastomycosis and phaeohyphomycosis inherently virulent?

Are these agents thermally dimorphic?

20) What are mycetomas caused by?

Which area(s) of the human body are commonly affected?

21) What area(s) of the body are affected by dermatophytoses?

22) List 3 dermatophyte genera and the diseases they cause.

What are the natural reservoirs for dermatophytes and how are they spread?

23) Where may ringworm affect the human body?

Why is a fungus called ringworm, when it isn't a worm?

How is ringworm diagnosed and treated?

24) How do superficial mycoses differ from subcutaneous mycoses?

Which are inflammatory infectious processes and which simply cause cosmetic problems?

25) What are black and white piedras? What parts of the human body are affected?

26) What diseases are caused by *Candida albicans*?

How do candidal infections start?

How may they be transmitted?

How is candidiasis diagnosed and treated?

27) What are cryptococcal infections?

What body systems are commonly affected by cryptococcal infection?

How is *C. neoformans* treated?

Who is at risk for C. neoformans infection?

What are the most common symptoms of systemic cryptococcosis?

28) How is cryptococcosis diagnosed and treated?

29) What is *Pneumocystis carinii* infection?

Why is the number of cases of this infection on the rise?

How is *Pneumocystis carinii* infection spread?

How is it diagnosed and treated?

30) What is aspergillosis and how is it transmitted?

How is aspergillosis diagnosed and treated?

31) What are zygomycoses caused by?

What pre-disposing conditions exist with most zygomycoses?

32) What diseases are associated with fungal allergens?

Organizing Your Knowledge

Fungus	Disease Caused	Type of infection	Reservoir & Endemic Areas	Portal of Entry	At risk populations	Diagnostic methods	Common Treatments
Histoplasma capsulatum							
Coccidiodes immitis							
	Chicago disease						
Paracoccidioides brasiliensis							
	Rose-gardener's disease						
	Tinea						
Pneumocystis carnii							
Candidia albicans							
Cryptococcus neoformans							

Testing Your Knowledge

1. Which of the following is NOT a common target of anti-fungal drugs?
    a. peptidoglycan
    b. plasma membranes
    c. cell division
    d. nucleic acid synthesis

2. The causative agent of California disease or San Joaquin Valley Fever is a(n):
    a. *Histoplasma*
    b. *Coccidioides*
    c. form of ringworm
    d. *Blastomyces*

3. Most fungal species are:
    a. true pathogens and readily infect humans
    b. opportunistic pathogens and may infect humans
    c. non-pathogenic to humans
    d. procaryotic

4. *Cryptococcus neoformans* infection generally leads to __.
    a. loss of hair
    b. lost of fingernails
    c. meningitis
    d. pulmonary infection

5. Which of the following is a subcutaneous fungal infection?
    a. blastomycosis
    b. sporotrichosis
    c. dermatophytosis
    d. histoplasmosis

6. Dermatophytes and Candida are both
    a. non-transmissible fungi
    b. lethal respiratory pathogens
    c. communicable fungi
    d. types of ringworm

7. Dimorphic fungi are generally ____ that ___.
    a. decomposers : live in the soil
    b. pathogens : cause disease
    c. commensual organisms: form spores
    d. non-pathogens : are extinct

8. Pathogenic fungi are generally ___ at 30 C and ____ at human body temperatures.
    a. procaryotic : eucaryotic
    b. molds : yeasts
    c. yeasts: molds
    d. haploid : diploid

9. Histoplasmosis outbreaks are associated with exposure to:
    a. infected people
    b. bird droppings
    c. contaminated food
    d. rose bushes

10. When rapid identification of a fungus is required, which technique is NOT generally used?
    a. serology
    b. PCR
    c. hyphal morphology
    d. culturing fungal spores

Chapter 23 The Parasites of Medical Importance

Building Your Knowledge

1) According to the World Health Organization (WHO), what percentage of all infections are caused by parasites?

2) What three factors have increased the worldwide distribution of parasites in recent years?

    a. _____

    b. _____

    c. _____

3) Why are some parasites that were once rare emerging as deadly pathogens?

4) How may parasites be spread from one host to another?

    a. _____

    b. _____

    c. _____

5) What are the 4 groups of protozoans?

    a. _____

    b. _____

    c. _____

    d. _____

6) How does a trophozoite differ from a cyst? Which is most often the infective body?

7) What is a karyosome and which stage of *Entamoeba histolytica's* life cycle is it found in?

8) What are chromatoidals and which stage of E. histolytica's life cycle are they found in?

What is the primary host for *E. histolytica*?

Where is the incidence of E. histolytica infection the highest?

What three variables determine the severity of infection?

    a. _____

    b. _____

    c. _____

How does *E. histolytica* cause disease in humans and what may happen in the course of untreated amebiasis?

How is amebic dysentery diagnosed? Are culture techniques commonly used to identify the organism? Why or why not?

9) Draw the *E. histolytica* life cycle. Label the cyst stage, encystment, and trophozoite stage. Indicate where trophozoites attach and multiply in the human body.

How can infection with *E. histolytica* be prevented?

10) What are two causative agents of amebic brain infection?

    a. _____

    b. _____

Where are these organisms normally found?

What outbreaks have occurred in the past few years?

Why is early treatment of *Naeleria fowleri* infection crucial?

11) What type of protozoan is *Balantidium coli* and what disease does it cause in humans?

How is balantidiosis commonly spread ?

Who is at risk of infection?

12) List four flagellated protozoans and give examples of the diseases they cause.

Flagellated Protozoan	Disease(s) Caused

13) What disease is caused by *Trichomonas vaginalis* infection and how is it spread?

How common is trichomoniasis in the United States?

How is the disease diagnosed and treated?

14) Which trichomonas species is associated with gum disease?

15) What causes giardiasis and what are the major symptoms of the disease?

How is *Giardia* transmitted and who is at risk of infection?

How is giardiasis diagnosed and treated?

How can giardiasis be prevented?

16) Why are trypanosomes and *Leishmania* species considered hemoflagellates?

How are hemoflagellates transmitted?

17) Draw the life cycle of *Trypanosoma cruzi*. Label the vector, human host and the amastigote, promastigote, epimastigote and trypomastigote stages.

Which stage is cyst-like?_____

Which stage is the most complex? _____

Does most differentiation take place in the vector or human host? _____

18) What disease is caused by *Trypanosoma brucei*?

Where is this disease endemic?

How does is this disease spread?

What are the major symptoms of *T. brucei* infection and how is it treated?

Why is there no effective vaccine against trypanosome infection?

19) What disease is cause by *Trypanosoma cruzi*?

How is it spread and where is T. cruzi infection endemic?

What is the prognosis for someone with untreated Chaga's disease?

20) How is leishmaniasis transmitted?

Which stage of the parasite's life cycle is transmitted to a human host?

Where do Leishmania parasites convert to the amastigote stage?

Differentiate between cutaneous leishmaniasis and visceral leishmaniasis. Which is the more serious illness?

Why is the diagnosis of leishmaniasis difficult?

How is leishmaniasis treated?

21) Which 3 human pathogens are apicomplexans?

a. _____

b. _____

c. _____

22) Which four species collectively cause malaria?

What percentage of the world's population lives in malaria-endemic regions?

How many new cases of malaria are diagnosed each year?

How many deaths are caused by malaria each year?

23) The life cycle of the malaria parasites has a sexual phase, carried out in the _____ and an asexual phase, carried out in the _____

Sporozoites enter liver cells, differentiate and reproduce. How many merozoites are produced for every infected liver cell? _____

Which stage of the parasite's life cycle is found in red blood cells?

How do malarial parasites complete sexual reproduction?

What are the symptoms and stages of malaria?

Why are relapses seen in certain types of malarial infection?

How does sickle cell anemia protect against malaria?

How is malaria treated?

How may humans reduce the risk of becoming infected with malaria?

Why is developing an anti-malarial vaccine so difficult?

24) What are the coccidian parasites?

How are they transmitted to humans?

25) What is *Toxoplasma gondii* and how widely distributed is the parasite?

What is the primary reservoir for toxoplasmosis?

How does the reservoir acquire the parasite?

Which populations are particularly at risk to develop severe complications from toxoplasmosis?

How is toxoplasmosis diagnosed and treated?

How can the disease be prevented?

26) What is sarcocystosis ?

How is it transmitted?

27) What disease does *Cryptosporidium* cause?

How is *Cryptosporidium* transmitted?

How is infection with *Cryptosporidium* diagnosed and treated?

28) How is *Cyclospora cayetanensis* transmitted?

What are the symptoms of infection?

How is it diagnosed and treated?

29) What organism causes redwater fever?

What are the symptoms of this disease?

30) How common is babesiosis in humans?

The research on babesiosis was ground breaking in two ways. Name these 2 ways.

a. _____

b. _____

31) What are hermaphroditic worms?

32) Differentiate between intermediate, definitive and transport hosts.

33) In most cases, how are humans exposed to helminth parasites?

What is the route of entry after exposure?

34) Why is the presence of parasites in or on humans more accurately considered an infestation rather than an infection?

35) In what areas of the world do most helminth diseases occur?

How do certain cultural practices aid in the spread of helminth populations?

Are most helminth infections localized or systemic?

How do ingested parasites invade a human host?

How do skin-burrowing pathogens spread throughout a host body?

36) Why do intestinal worms often cause weight loss?

37) Which type of leukocyte is most responsible for eliminating worms?

38) How are helminth infections diagnosed?

Why are most anti-helminth drugs toxic to the host as well as the parasite?

39) How are intestinal nematodes different from tissue nematodes? Give examples of each.

40) How is ascariasis transmitted from one host to another?

What happens to the nematode after a human ingests *Ascaris* eggs?

41) Differentiate between pinworm and whipworm infections. Which is more serious?

42) What is the "hook" of a hookworm?

How are hookworms different from the other intestinal parasites discussed so far?

43) What is a threadworm infection and what helminth infection does it resemble?

Who is at risk for serious complications with threadworm infections?

44) How do humans catch trichinosis?

Why are humans dead-end hosts to the infectious cycle?

45) Name three filarial worms and the diseases they cause.

Filarial Worm	Disease Caused

Are filarial worms tissue or intestinal parasites?

46) What is the causative agent of elephantiasis?

What are the symptoms of elephantiasis and how are these symptoms related to the growth of the parasite in the human body?

47) What vector transmits river blindness?

How may antibiotics, such as tetracycline, be helpful in treating a parasitic disease, such as river blindness?

48) Schistosomes have a complex life cycle with many stages. What stage infects snails?

Which stage infects humans? _____

How widespread is schistosomiasis, worldwide?

What are the stages and symptoms of schistosomiasis?

What steps are being taken to control the spread of schistosomiasis?

49) How do lung and liver flukes differ from one another?

50) Tapeworms have a simple structure of scolex and proglottids. Please draw a tapeworm, labeling the scolex and proglottids and indicate what each is used for during the life cycle of the parasite.

51) Describe three separate types of tapeworm infection.

Tapeworm	Diagnostic methods	Treatment

Organizing Your Knowledge

Parasitic Agent	Nature of Parasite	Disease caused	Symptoms
Ascaris lumbricoides			
Balantidium coli			
Cryptosporidium			
Cyclospora			
Entamoeba histolytica			
Giardia lamblia			
Leishmania tropica			
Naegleria fowleri			
Necator americanus			
Onchocerca volvulus			
Plasmodium spp.			
Schistosoma spp.			
Taenia spp.			
Trichinella spiralis			
Trichomonas tenax			
Trichomonas vaginalis			
Trichuris trichiura			
Trypanosoma brucei			
Trypanosoma cruzi			
Wurchereria bancrofti			

Parasitic Agent	Diagnostic Methods	Transmitted By____	Prevention
Ascaris lumbricoides			
Balantidium coli			
Cryptosporidium			
Cyclospora			
Entamoeba histolytica			
Giardia lamblia			
Leishmania tropica			
Naegleria fowleri			
Necator americanus			
Onchocerca volvulus			
Plasmodium spp.			
Schistosoma spp.			
Taenia spp.			
Trichinella spiralis			
Trichomonas tenax			
Trichomonas vaginalis			
Trichuris trichiura			
Trypanosoma brucei			
Trypanosoma cruzi			
Wurchereria bancrofti			

Practicing Your Knowledge

1. The active feeding state of a typical protozoan parasite is a _____
    a. cyst
    b. hemoflagellate
    c. phlebotomine
    d. trophozoite

2. The least common protozoan infections are caused by ___.
    a. apicomplexans
    b. ciliates
    c. ameobas
    d. flagellates

3. Which of the following is NOT a pathogenic protozoan group?
    a. ciliates
    b. diatoms
    c. flagellates
    d. apicomplexans

4. Roundworms and tapeworms are both ___.
    a. ciliate protozoans
    b. helminths
    c. vector-borne parasites
    d. filarial worms

5. Arthropods are transmission vectors for all of the following EXCEPT __.
    a. malaria
    b. Chaga's disease
    c. Leishmaniasis
    d. trichinosis

6. The female Anopheles mosquito completes the ___ phase of the malarial parasite's life cycle.
    a. asexual
    b. sexual
    c. merozoite
    d. hemolytic

7. Why are most helminth diseases most accurately described as infestations, not infections?
    a. all helminths are hermaphroditic
    b. all helminths reproduce asexually
    c. adult helminths don't multiply and grow to maturity in a single host
    d. no helminths can inflitrate the blood

8. Water-borne parasites include ___.
    a. *Plasmodium* and *Trichomonas*
    b. *Giardia* and *Cryptosporidium*
    c. promastigotes and amastigotes
    d. *Legionella* and *Leishmania*

9. Schistosomiasis and malaria are both___.
    a. caused by protozoan infections
    b. gastrointestinal illnesses
    c. blood diseases with liver involvement
    d. transmitted by mosquito bite

10. The causative agents for elephantiasis and schistosomiasis are both ___.
    a. filarial worms
    b. blood flukes
    c. mosquito-borne
    d. parasites

11. Prevention of toxoplasmosis is best accomplished by ___.
    a. use of insect repellant
    b. avoiding snail-infested waters
    c. wearing shoes when walking
    d. proper hygiene around cats
    e. not eating raw pork

12. A protozoan flagellate disease that is sexually transmitted is ___.
    a. Giardiasis
    b. Leishmaniasis
    c. Trichomoniasis
    d. Trypanosomiasis

13. Ameboid disease may directly cause ____ in humans.
    a. dysentery and liver failure
    b. muscle fatigue and anemia
    c. sleeping sickness
    d. meningoencephalitis and dysentery

14. Which of the following is mis-matched?
    a. *Ascaris* - roundworm
    b. Amoeba - *Naegleria*
    c. Trypanosoma - helminth
    d. malaria - apicomplexan

15. The causative agent of malaria is ___.
    a. an apicomplexan
    b. a filarial worm
    c. a hemoflagellate
    d. a neurociliate

Chapter 24 Introduction to the Viruses of Medical Importance:

The DNA viruses

Building Your Knowledge

1) How are viruses significantly different from bacteria, fungi and protozoans?

2) How are animal viruses divided into families?

   Are most DNA viruses generally single-stranded or double-stranded? Are most RNA viruses single or double stranded?

   Where does the envelope of an animal cell virus come from?

3) Why are certain viruses associated with certain cell types (for example, hepatitis and liver cells)?

4) How do humans develop immunity to viruses?

Why are vaccines often the most effective defense against viral infection?

5) What three viruses are known for their tetragenic effects?

    a. _____

    b. _____

    c. _____

6) What are the 6 groups of DNA viruses that are pathogenic to humans?

    a. _____

    b. _____

    c. _____

    d. _____

    e. _____

    f. _____

7) Describe the poxviruses. Where do these poxviruses proliferate in the human body? What causes pox to form?

8) What is the difference between variola and vaccinia?

9) How serious is smallpox infection? How was it eradicated and when?

Why are health officials considering re-starting vaccination programs in the United States?

What are the drawbacks to a modern wide-spread smallpox vaccination program?

How is smallpox transmitted?

What are the symptoms of smallpox?

Which virus is commonly used to vaccinate against smallpox?

10) What is molluscum contagiosum and how is it contracted?

How is it treated?

11) Which mammalian poxviruses can cause disease in humans?

12) List the pathogenic herpesviruses and the diseases they cause.

13) Compare and contrast herpes simplex type 1 and type 2 infections.

How are the 2 viruses spread?

What are the symptoms and complications of infection?

How do each become latent and how do recurrent attacks occur?

How does herpes simplex 1 manifest itself in children?

How is herpes simplex 2 transmitted and what are the major symptoms of the disease?

How are herpes infections diagnosed and treated?

What is whitlow and who is in danger of contracting it?

14) Which virus is the causative agent of chicken pox?

Which virus causes shingles?

How was a connection made between the two diseases?

Can someone catch chickenpox from someone with shingles?

What pattern do the lesions associated with shingles follow?

How are shingles activated?

How can shingles and chicken pox be treated?

15) What is CMV?

How widespread is CMV infection and how is CMV transmitted ?

What are the symptoms of congenital CMV?

16) What are the symptoms of perinatal CMV?

What is disseminated cytomegalovirus infection?

Who is at risk for disseminated CMV?

How is it treated?

What problems have been encountered to develop a vaccine for CMV?

17) How are Burkitt's lymphoma and mononucleosis related?

18) How does EBV infection manifest itself in developing countries?

How does EBV infection manifest itself in industrialized countries?

Why is there a difference?

19) Describe three separate illnesses caused by EBV.

    a. _____

    b. _____

    c. _____

20) How is EBV diagnosed, treated and prevented?

21) What diseases are caused by HHV-6?

22) If cancer patients are sero-positive for herpesviruses, does that indicate the virus caused the cancer? Why or why not?

23) What organ do hepadnaviruses infect? What disease(s) do they cause?

24) How many different viruses can cause hepatitis?

Are these viruses related to one another?

25) How is hepatitis B spread?

What is the infectious dose?

What is the range of HBV symptoms?
How is hepatitis B diagnosed?

Can it be treated? If so, how is it treated?

How can hepatitis B virus infection be prevented?

26) What diseases are associated with adenovirus infection?

How are adenoviruses transmitted from person to person?

27) What diseases are caused by papillomaviruses?

Which of these are sexually transmitted?

What are the sequelae of infection?

28) What are the polymaviruses?

What diseases are they associated with?

29) What one group of pathogenic DNA viruses are single-stranded?

What diseases do these viruses cause in humans?

What diseases do these viruses cause in animals?

What are the congenital effects of infection?

Organizing Your Knowledge

Virus	Describe Virus	Disease Caused & Diagnostic methods	Prevention & Treatment
Adenovirus			
Cytomegalovirus			
Epstein-Barr			
HBV			
HSV I			
HSV II			
Papillomavirus			
Parvovirus			
Vaccinia			
Varicella-zoster			
Variola			

Virus	Disease Caused	Tissue Tropism	Transmitted by ___.
Adenovirus			
Cytomegalovirus			
Epstein-Barr			
HBV			
HSV I			
HSV II			
Papillomavirus			
Parvovirus			
Vaccinia			
Varicella-zoster			
Variola			

Practicing Your Knowledge

1. Which of the following is NOT a sexually transmitted disease?
   a. papilloma
   b. variola
   c. hepatitis C
   d. all of the above are transmitted primarily through sexual contact

2. Adenoviruses are ___,
   a. single-stranded RNA viruses
   b. double-stranded DNA viruses
   c. single-stranded DNA viruses
   d. they are single-stranded DNA viruses

3. Vaccinia and variola are both ___.
   a. poxviruses
   b. parvoviruses
   c. herpesviruses
   d. adenoviruses

4. Most DNA viruses are assembled in the _____ of an animal cell.
   a. cytoplasm
   b. Golgi apparatus
   c. mitochondria
   d. nucleus

5. Parvoviruses are unusual because they ____.
   a. cause cold-like symptoms
   b. are double-stranded DNA viruses
   c. have both DNA and RNA in the same viral particle
   d. are single-stranded DNA viruses
   e. infect only humans

6. Tetragenic viruses cause ___ and include ___.
   a. cancer : smallpox
   b. heart attacks : HIV
   c. birth defects : rubella
   d. meningitis: meningococcus

7. The Epstein-Barr virus causes ____.
   a. shingles
   b. birth defects
   c. hepatitis
   d. mononucleosis

8. Viral hepatitis occurs ___.
   a. as a result of many different viral infections
   b. as the result of single viral infection
   c. only in humans
   d. only in infants

9. The smallpox virus reproduces in ___ cells primarily.
   a. liver
   b. epidermal
   c. nerve
   d. muscle

10. Most herpes viruses cause ___.
   a. cold sores to form
   b. latent viral infections that can recur
   c. AIDS
   d. cold and flu-like symptoms

11. Chickenpox is caused by a ____.
   a. adenovirus
   b. poxvirus
   c. herpesvirus
   d. cytomegalovirus

12. Most DNA viruses that are human pathogens are ___.
   a. single stranded
   b. double stranded
   c. there are no DNA viruses that are human pathogens
   d. triplex

Chapter 25 The RNA Viruses of Medical Importance

Building Your Knowledge

1) How many groups of RNA viruses cause disease in humans?

   How are these viruses separated?

2) How does a segmented viral genome differ from a non-segmented genome?

3) Draw an orthomyxovirus. Label the hemagglutinin and neuraminidase, envelope and RNA molecule.

4) How do hemagglutinin and neuraminidase contribute to the virulence of influenza virus?

5) Differentiate between antigenic shift and antigenic drift.

Would influenza be as capable of antigenic shift if it had a non-segmented genome? Why or why not?

Why is it necessary to get a flu shot every year?

6) How are flu viruses named?

Which flu pandemic of the 20th century had the greatest world-wide death toll?

What anti-viral drugs are available to treat influenza?

7) What disease is caused by hantaviruses and how is it spread?

8) What are the 3 major paramyxoviruses and what diseases do they cause?

Paramyxovirus	Disease caused

How are all three paramyxoviruses spread?

9) What are the symptoms of parainfluenza virus infection in children and how are they treated?

Who is most susceptible to infection with parainfluenza virus?

10) How is the mumps virus transmitted?

Where does the virus multiply early in an infection?

Where does the virus multiply late in the infection?

How many different serotypes are there of the mumps virus?

How does the number of different serotypes affect the usefulness of a vaccine?

How common is mumps-related sterility in males?

How could a respiratory pathogen be associated with reproductive capacity?

How is mumps diagnosed and treated?

11) What is the natural reservoir for measles?

12) How is measles infection spread?

When are people who contract measles contagious?

13)How is measles diagnosed and treated?

How is measles infection prevented?

14)What are the symptoms of RSV in adults and older children?

What are the symptoms of RSV in infants?

Who is at the most risk to develop serious complications from RSV infection?

15)Which rhabdovirus is most concerning to human medicine?

What is the prognosis for untreated rabies?

How do humans contract rabies?

Where does rabies multiply early in an infection?

Where is the virus found in late stages of an infection?

Compare and contrast the furious and dumb forms of rabies.

Why is the diagnosis of rabies so difficult?

Once a rabies has been diagnosed, what is the typical course of treatment?

How is this unusual?

How is rabies infection prevented?

16)Describe a coronavirus.

What diseases are caused by coronaviruses?

17) Describe a togavirus.

What viral groups are togaviruses?

18) What is rubella and how is it spread?

Describe the two clinical forms of rubella.

Which is more serious?

How is rubella diagnosed?

How is rubella prevented?

19) How are arboviruses transmitted?

What are the groups of arboviruses that are pathogenic to humans?

20) What factors influence the distribution and frequency of arboviral infections?

Which seasons have the greatest number of arboviral infections? Why?

21) What are the arboviruses that cause encephalitis?

Are humans the primary host for these viral pathogens?

22) What do yellow fever and Dengue fever have in common?

How is yellow fever spread?

Differentiate between urban and sylvan spread of yellow fever.

23) How are arboviruses diagnosed, treated or prevented?

24) How do retroviruses differ from most RNA viruses?

What diseases do the known retroviruses cause ?

25) Draw an HIV viral particle, labeling the envelope, spikes and RNA.

26) How and when was AIDS first characterized in the United States?

Where and when was the first documented case of AIDS?

How many people are infected with HIV world-wide?

27) What does the HIV blood screening test measure?

28) Describe the HIV infection and AIDS case classification system. How do A categories differ from B categories? How do B categories differ from C?

29) Describe the seven risk categories for developing HIV infection, starting with the most prevalent.

Why have congenital HIV infections dramatically decreased in recent years?

30) How is HIV transmitted?

Are fomites a common part of the HIV infection cycle?

Are mosquitoes carriers of HIV?

31) Diagram the infection of HIV in the blood.

Which cells are infected? What happens to viruses when they enter host cells?

32) How and why do T cells die during an HIV infection?

How do HIV viral particles enter the brain?

33) Which opportunistic infections generally mark the transition from HIV-positive to AIDS?

What cancers do AIDS patients commonly have?

How can a person get a false-negative HIV test?

34) Please describe four separate anti-AIDS drugs and the viral life cycle segment they target.

Reverse transcriptase inhibitors -

Protease inhibitors -

Anti-integrase drugs -

Fusion inhibitors –

35) How may HIV transmission be prevented?

36) What are the challenges to developing an HIV vaccine?

37) Name two different retroviral agents and the diseases they cause.

38) What type of virus causes polio?

39) How is polio transmitted?

Why is it significant that polio viral particles are resistant to acid and bile?

How does infection with a polio virus progress?

What is post-polio syndrome?

40) Differentiate between the Salk and Sabin vaccines. Which is live?   Which is an inactivated vaccine? What are the advantages and disadvantages of each?

Vaccine	Nature of Vaccine	Advantages	Disadvantages
Salk vaccine			
Sabin vaccine			

41) What are the non-polio enteroviruses and close relatives?

42) What is the common-cold syndrome and what are its symptoms?

43) What is hepatitis A?

How is it spread?

What are the symptoms of hepatitis A?

44) Why are rhinoviruses restricted to growth in the nose and upper respiratory areas?

Why is developing a vaccine against the common cold highly unlikely?

How may people minimize the spread of colds?

45) How are rotavirus and Norwalk virus similar?

How are they different?

Who typically gets rotavirus? How is severe rotavirus infection treated?

46) How is a prion different from a viral particle?

Name three diseases known to be caused by prions?

How does prion infection cause disease?

How does one contract variant CJD?

Organizing Your Knowledge

Virus	Viral Group	Virus Description	Disease Caused & Symptoms
Adult T cell leukemia			
Common cold			
Dengue fever			
Hepatitis A			
HIV			
influenza virus			
Measles			
Mumps			
Parainfluenza			
Polio			
Rabies			
Rabies			
Rotavirus			
RSV			
Rubella			
SARS			
Sin Nombre			
West Nile			
Yellow fever			

Virus	Disease	Transmitted by ___	Prevention and treatment
Adult T cell leukemia			
Common cold			
Dengue fever			
Hepatitis A			
HIV			
influenza virus			
Measles			
Mumps			
Parainfluenza			
Polio			
Rabies			
Rabies			
Rotavirus			
RSV			
Rubella			
SARS			
Sin Nombre			
West Nile			
Yellow fever			

Practicing Your Knowledge

1. Arboviruses are all____.
   a. zoonotic diseases
   b. arthropod-borne
   c. RNA viruses
   d. all of the above
   e. none of the above

2. In nature, Yellow fever and Dengue Fever are primarily transmitted by ____.
   a. mosquitos
   b. air droplets
   c. blood products
   d. unsafe sexual practices

3. How is poliovirus spread throughout a population?
   a. air-borne routes
   b. BSE-contaminated meat
   c. fecal-oral contamination
   d. blood and other bodily fluids

4. HIV is the only retrovirus that infects humans.
   a. True
   b. False

5. Mumps, polio and rubella are all ____.
   a. RNA viruses
   b. zoonotic diseases
   c. spread through fecal-oral transmission
   d. enveloped viruses

6. The difference between HIV infection and AIDS is
   a. there is a different causative agent
   b. AIDS patients are positive for anti-HIV antibodies, HIV patients are not
   c. AIDS patients chave circulating virus, HIV patients do not
   d. AIDS is associated with decreased T cell counts and indicator conditions

7. Most RNA viruses are ____ and assemble in the __.
   a. double stranded : nucleus
   b. single stranded : nucleus
   c. double stranded : cytoplasm
   d. single stranded : cytoplasm

8. Which of the following is not a disease commonly found in AIDS patients?
   a. *Pneumoncystis carini*
   b. Kaposi sarcoma
   c. kuru
   d. *Cryptococcus* pneumonia

9. If influenza had a non-segmented genome, it would __.
   a. be related to the prion diseases
   b. not be infectious
   c. be less likely to show antigenic drift
   d. be a DNA virus

10. Hepatitis A is caused by ____
    a. a retrovirus
    b. a poliovirus
    c. a picornavirus
    d. a togavirus

11. HIV anti-viral drugs target __
    a. proteases
    b. reverse transcriptase
    c. the fusion of viral particles with cells
    d. all of the above

12. All retroviruses produce _
    a. hemagglutinin
    b. parainfluenza
    c. reverse transcriptase
    d. spongiform particles

13. The causative agent of Severe Acute Respiratory Syndrome (SARS) is _.
a. an influenza virus
b. a pneumovirus
c. a rhabdovirus
d. a coronavirus

14. Rabies is an unsual infection because it:
a. is untreatable
b. responds to active immunization after exposure
c. can't be spread between species
d. is a viral infection that responds to antibiotics

15. How may HIV be transmitted?
a. mosquito bites
b. handshakes
c. blood and bodily fluids
d. all of the above

Chapter 26 Environmental and Applied Microbiology

Building Your Knowledge
1) Differentiate between applied and environmental microbiology.

2) Draw the levels of the biosphere as they relate to one another. Label the biosphere, hydrosphere, lithosphere and atmosphere. There will be overlap.

3) How is a community different from a population?

4) How does an organism's niche differ from its habitat?

Which organism would likely have a broader niche, a scavenger or a nitrogen fixer? Explain your answer.

5) Are producers autotrophic or heterotrophic?

What do producers produce and how do they produce it?

What role do consumers play in an ecosystem?

What would happen to an ecosystem that lost its decomposers?

6) Why does the energy available to consumers decrease as you move from primary to secondary and from secondary to tertiary consumers?

7) Why is a food web a better model than a food chain, when describing the trophic relationships in an ecosystem?

8) Define and give examples of the ecological relationships listed.

Relationship	Definition	Example
Synergism		
Parasitism		
Predation		
Competition		

9) What 5 Traits do most biogeochemical cycles share?

a. _____

b. _____

c. _____

d. _____

e. _____

10) What do giomicrobiologists study?

11) Which atmospheric cycle is most closely associated with living systems, energy transformation, and trophic patterns?

12) What process fixes carbon dioxide into an organic form of carbon (sugar)?

How is carbon dioxide released from living processes?

What gas do methanogens release?

13) Write the basic equation for photosynthesis.

What happens during the light-dependent reactions of photosynthesis?

14) Why are photosynthetic pigments crucial to photosynthesis?

What happens during the light-independent reactions of photosynthesis?

15) How is ATP synthesized during photosynthesis and when is it synthesized?

What happens to the ATP that is synthesized?

How is carbon fixed?

16) Compare and contrast oxygenic and anoxygenic photosynthesis?

Which is most common?

17) Diagram the process of nitrogen fixation.

What organisms are able to fix nitrogen? Are plants able to fix nitrogen?

How is ammonia produced from amino acids?

What are the nitrifying bacteria?

What are the detrifying bacteria?

18) How is sulfur used in living organisms?

How is the sulfur cycle similar to the nitrogen cycle?

How does organic sulfur differ from inorganic?

19) How can *Thiobacillus* survive in environments with few complex organic nutrients?

What role does *Thiobacillus* play in the phosphorus cycle?

What role does phosphorus have in living organisms?

Why is phosphorus a major component of fertilizers?

What effect does excess phosphorus have on the hydrosphere?

20) What is bioamplification and how do microbes play a role in the process?

21) How does soil composition determine the oxygen and carbon dioxide concentration of the lithosphere?

How does soil composition influence the distribution of organisms in the soil?

22) What is humus?

How does humus production and use change with different environments?

Bogs and tropical forests both have nutrient poor soil, but for different reasons. Why are tropical forest soils generally poor?

Why are bog soils generally nutrient-deficient?

23) What is the rhizosphere?

24) How are mycorrhizae beneficial to plants?

25) What waste compounds are common soil contaminants?

26) Draw a hydrologic cycle, labeling transpiration, precipitation, aquifers, surface water and clouds.

27) What two problems are acquirers currently facing?

28) Draw a lake in crows-section. Label the photic, profundal, and benthic zones. Label the littoral and limnetic zones as well.

29) What types of organisms are commonly found in estuaries?

30) Differentiate between phytoplankton and zooplankton. Which zone of a lake or ocean do both inhabit?

31) What is thermal stratification?

32) Which bacterial species tend to live in oligotrophic lakes?

What adaptations do they need to live there?

33) What role do bacteriophage play in aquatic environments?

34) What are the most prominent human pathogens that are water-borne?

35) What 3 traits do indicator bacteria need to have in order to be useful at indicating fecal contamination of water?

        a. _____

        b. _____

        c. _____

36) Why are coliforms good indicatororganisms but coliform bacteriophages are not?

37) Compare and contrast standard plate counts with membrane filter methods.

Which is most useful for larger volumes of water?

38) Why is it important to distinguish between fecal and non-fecal coliforms?

What is an acceptable level of fecal coliforms in drinking water?

39) How is drinking water purified? Diagram the process from water source to kitchen sink.

40) How is sewage purified and what role do microbes play in the process? Diagram the process from kitchen sink to waterways.

41) What are the earliest examples of humans using microbes to do work?

42) How do industrial microbiologists define fermentation?

43) Give two examples of beneficial actions of microbes on food and detrimental actions of microbes on food.

44) Do the same microbes that spoil food cause food-borne infections? Explain.

45) How is making bread similar to the process of making beer? How is it different?

46) Diagram the process of beer-making from ingredients to finished product.

47) Why is most beer between 3% and 5% alcohol?

48) Diagram the process of winemaking from grape harvest to bottling.

Why is the maximum alcohol content of wine around 17%?

49) Fill in the following chart.

Item fermented	Product
Potatoes	
	Whiskey
Corn Mash	
	Brandy

50) How is vinegar made?

51) How is cheese made? Where does rennin come from and what is it used for?

What factors give different cheeses different tastes?

Name 3 other fermented milk products.

52) How does a food-borne infection differ from a food-borne intoxication?

53) How are the food-borne toxins named?

54) How many food-borne illnesses occur each year in the United States?

55) How is the incidence of food-borne infection reduced in food processing and preparation?

56) How does UHT pasteurization differ from HTST pasteurization?

Which process leads to a product with a longer shelf life?

57) Why can utensils be treated with UV radiation, but food is treated with gamma irradiation?

Does irradiated food become radioactive?

58) What is the mechanism of action and the food the following preservatives are commonly used in?

Preservative	Mechanism of Action	Example
Nitrites and Nitrates		
Organic Acids		
Sugar		
Salt		

59) How do primary and secondary metabolites differ from one another?

60) Why are most strains of bacteria used in industry genetically altered?

61) How may industrial microbiologists increase the amounts of a desired end product?

62) Describe the process of growing organic substances in a fermentor and the mass production of those substances.

63) How does a continuous system differ from batch fermentation?

64) What was the first mass-produced antibiotic?

65) What other substances of pharmaceutical value are produced this way?

Organizing Your Knowledge

Please indicate with an X in the appropriate column(s), which branch of microbiology each area of study belongs to.

Area of Study	Applied Microbiology	Environmental Microbiology
Bioremediation		
Water Microbiology		
Food Microbiology		
Energy Pyramids		
Trophic structures		
Food webs		
Biogeochemical cycles		
Biomagnification		
Sewage Treatment		
Production of Enzymes		
Mass Production of Drugs		
Production of Vaccines		
Chemical cycling		

Cycle	Organic Forms	Fixers	Inorganic Forms	Decomposers
Carbon				
Nitrogen				
Sulfur				
Phosphorus				

Practicing Your Knowledge

1. The same organisms that spoil food will often cause disease in humans.
   - a. True
   - b. False

2. Rennin is an important part of ___.
   - a. industrial fermentors used to make penicillin
   - b. bread-making
   - c. beer making
   - d. cheese-making

3. Whey is a _____.
   - a. normal product of yogurt culture
   - b. byproduct of wine-making
   - c. toxic metabolite from cheese-making
   - d. normal product of cheese production

4. In photosynthesis, the ATP that is produced _____.
   a. is used to fix carbon dioxide to sugar
   b. is used by the cell
   c. powers photosystem II
   d. powers photosystem I

5. Which of the following is NOT a commonly used method to preserve food?
   a. radiation
   b. vinegar
   c. salt
   d. milk proteins

6. Plankton often live in the _____ zone of a lake
   a. benthic
   b. profundal
   c. abyssal
   d. photic

7. Different soil levels differ greatly on their _____.
   a. oxygen available
   b. microbial flora
   c. acidity
   d. all of the above

8. Microbes are a crucial part of the _____ stage of sewage treatment.
   a. primary
   b. secondary
   c. tertiary
   d. chlorination

9. Primary metabolites are_____.
   a. generally what's harvested from a fermentor
   b. essential to a microbe's function
   c. generally toxic to humans
   d. are synthesized during log phase of a growth cycle

10. Biomagnification contributes to the -
    a. accumulation of decomposers in the soil
    b. accumulation of toxins in the food chain
    c. greenhouse effect
    d. depletion of ozone

11. The soil in bogs is generally nutrient-poor because ____.
    a. microbes decompose nutrients too quickly
    b. not enough plant life to form nutrients
    c. too much oxygen for anaerobes to work properly
    d. high acid content and slow microbial decomposition

12. Filtration of water for drinking is an important way to _____.
    a. remove bacteriophage
    b. remove protozoan cysts
    c. chemically disinfect water
    d. irradiate fresh water

13. If a person catches Salmonella from eating undercooked chicken, that person has a food-borne _____.
    a. infection
    b. intoxication
    c. spoilage
    d. fermentation

14. The form of pasteurization that will result in the longest shelf life is ____.
    a. high-temperature, short time
    b. low temperature, long time
    c. ultra-high temperature
    d. ultra-low temperature

15. Excessive phosphorus in run-off will often lead to ____.
    a. oligotrophication of lakes
    b. a spike in giardiasis cases
    c. eutrophication of lakes
    d. no deleterious effects - the environment absorbs the excess

Answer Key To "Practicing Your Knowledge" Section

Chapter 1	Chapter 2	Chapter 3	Chapter 4	Chapter 5	Chapter 6
1. E	1. A	1. D	1. A	1. C	1. A
2. B	2. E	2. C	2. C	2. B	2. B
3. E	3. E	3. D	3. D	3. B	3. A
4. A	4. D	4. C	4. C	4. B	4. B
5. A	5. D	5. A	5. C	5. A	5. D
6. C	6. C	6. D	6. D	6. A	6. C
7. B	7. A	7. D	7. B	7. C	7. B
8. D	8. D	8. A	8. C	8. C	8. A
9. E	9. C	9. C	9. B	9. C	9. C
10. D	10. D	10. E	10. A	10. B	10. A
11. D	11. B	11. B	11. A	11. D	11. C
12. C	12. B	12. B	12. B	12. C	12. A
13. B	13. D	13. B	13. B	13. D	13. D
14. C	14. E	14. B	14. A	14. A	14. C
15. A	15. C	15. E	15. B	15. A	15. B

Chapter 7	Chapter 8	Chapter 9	Chapter 10	Chapter 11	Chapter 12
1. C	1. B	1. A	1. C	1. A	1. D
2. D	2. A	2. C	2. D	2. B	2. B
3. D	3. D	3. D	3. B	3. D	3. D
4. B	4. B	4. C	4. C	4. B	4. A
5. D	5. B	5. B	5. B	5. B	5. B
6. C	6. B	6. A	6. D	6. C	6. B
7. B	7. C	7. D	7. B	7. A	7. B
8. B	8. D	8. D	8. C	8. B	8. D
9. C	9. C	9. C	9. A	9. C	9. C
10. B	10. B	10. C	10. B	10. C	10. C
11. A	11. B	11. D	11. D	11. A	11. C
12. B	12. D	12. C	12. A	12. C	12. A
13. B	13. D	13. B	13. D	13. D	13. C
14. D	14. C	14. B	14. C	14. B	14. A
15. A	15. D	15. C	15. C	15. A	15. C

Chapter 13	Chapter 14	Chapter 15	Chapter 16	Chapter 17	Chapter 18
1. D	1. D	1. C	1. D	1. A	1. C
2. C	2. A	2. A	2. B	2. C	2. C
3. D	3. B	3. D	3. D	3. A	3. C
4. C	4. B	4. B	4. D	4. B	4. D
5. B	5. D	5. A	5. A	5. D	5. A
6. B	6. C	6. D	6. C	6. A	6. B
7. A	7. C	7. D	7. A	7. C	7. C
8. D	8. C	8. C	8. C	8. D	8. D
9. C	9. B	9. B	9. D	9. B	9. B
10. A	10. C	10. A	10. C	10. B	10. A
11. C	11. D	11. C	11. A	11. B	
12. A	12. B	12. D	12. A	12. A	
13. C	13. A	13. D	13. B	13. D	
14. A	14. C	14. B	14. C	14. C	
15. B	15. D		15. D	15. C	

Chapter 19	Chapter 20	Chapter 21	Chapter 22	Chapter 23	Chapter 24
1. C	1. C	1. C	1. A	1. D	1. B
2. B	2. D	2. A	2. B	2. B	2. B
3. A	3. C	3. D	3. C	3. B	3. A
4. C	4. D	4. C	4. C	4. B	4. D
5. B	5. C	5. C	5. B	5. D	5. D
6. D	6. C	6. B	6. C	6. B	6. C
7. D	7. A	7. D	7. B	7. C	7. D
8. C	8. C	8. A	8. B	8. B	8. A
9. C	9. B	9. B	9. B	9. C	9. B
10. C	10. B	10. C	10. D	10. D	10. B
11. B	11. B	11. D		11. D	11. C
12. D	12. B	12. B		12. C	12. B
13. A	13. D	13. A		13. D	
14. D	14. C	14. C		14. C	
15. B	15. A	15. B		15. A	

Chapter 25

1. D
2. A
3. C
4. B
5. A
6. D
7. D
8. C
9. C
10. C
11. D
12. C
13. D
14. B
15. C

Chapter 26

1. D
2. B
3. D
4. C
5. B
6. A
7. D
8. B
9. B
10. C
11. D
12. D
13. D
14. A
15. B

Notes

Notes

Notes

Notes

Notes

Notes

Notes

Notes